EXHIBIT CATALOG OF THE DA VINCI EXPERIENCE

FIFTH EDITION

COMPILED AND EDITED BY
Godfrey Harris

THE AMERICAS GROUP

First Edition, 1st Printing — October 2006
Second Edition, 1st Printing — March 2007
Third Edition, 1st Printing — October 2007
Fourth Edition, 1st printing — January 2009
Fifth Edition, 1st Printing — January 2010

The Americas Group
654 N. Sepulveda Boulevard, Suite 1
Los Angeles, California 90049-2070
U.S.A.
☎ + (1) 310 476 6374
FX + (1) 310 471 3276
EM. hrmg@mac.com
www.AmericasGroup.com

ISBN:
978-0-935047-58-5

Library of Congress Cataloging-in-Publication Data

Library of Congress Cataloging-in-Publication Data

Harris, Godfrey, 1937-
Exhibit catalog of the Da Vinci experience / compiled and edited by Godfrey Harris.
p. cm.
Includes bibliographical references and index.
ISBN-13: 978-0-935047-58-5
1. Machinery--History--Exhibitions. 2. Leonardo, da Vinci, 1452-1519--Exhibitions. I. Title.

TJ17.H37 2006
621.8092--dc22

2006028130

Printed in India
Aegean Offset Printers

TABLE OF CONTENTS

ABOUT
LEONARDO DA VINCI

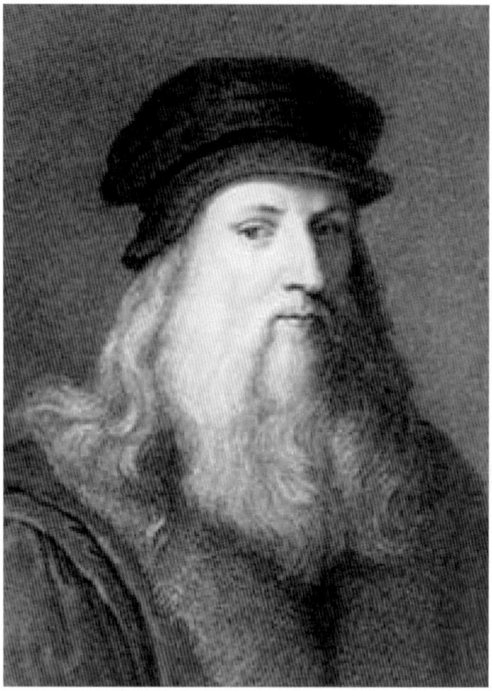

Leonardo da Vinci believed in learning all of his life and thought that "every difficulty [could] be overcome by effort."

ABOUT LEONARDO DA VINCI

Leonardo da Vinci is one of the most intriguing men in history— an ordinary man who did extraordinary things. He was much more than just an artist; he was also an architect, designer, engineer, philosopher, and scientist. He learned to use art in many ways — to tell a story, to capture a mood, to decorate a space, to illustrate a process, as well as to better explain anatomy, botany, and geology. Leonardo wanted to capture what he observed in nature to improve on what was known in his day and to develop new ideas.*

---

* The idea of drawing, as a means of recording activities and expressing ideas, has been made famous by Leonardo's more than 14,000 pages of sketches. Drawing was considered such an important skill that every educated child was expected to master it, much as we are now expected to learn reading and writing. Between 1820 and 1860, there were no fewer than 145,000 drawing manuals in circulation in the United States alone! While Benjamin Franklin called drawing "a universal language, understood by all nations," it started to fade in importance in the 1890s when Americans seemed to worry more about feelings and self-esteem than dexterity and discipline. This change may also have signaled changes in modern art where the rule books on draftsmanship were set aside for a new, free-thinking attitude. (Adapted from Michael Kimmelman's "The Art of Drawing," *The New York Times*, July 19, 2006.)

How did this one individual come to accomplish so many things in so many fields? Why this man at this point in human history? As a start to answering this question, the time of Leonardo's life must be understood in its historical context: Constantinople had just fallen to the Turks, creating a great exodus of scholars with their manuscripts, books, scrolls, and ideas to what is now northern Italy, the center of learning of the day; Vasco da Gama had sailed around Africa to India; Columbus had found a New World; great wealth, based on gold and food-preserving spices, flowed to support art — not only Leonardo's but that of Bottecelli, Raphael, Michelangelo, and a dozen other giants of the day.

Leonardo enters this intellectual and creative ferment with a desire to succeed that seems to be driven by forces greater than his insatiable curiosity and enormous imagination. One theory holds that his childhood was filled with such personal doubts and family uncertainties that he did everything in his power throughout his life to ensure that he would not be forgotten or ignored, that his work and effort would receive the respect and attention he sought and thought they deserved.

Leonardo, after all, had to face the fact that he was the illegitimate son of a prominent Florentine notary. At an early age he was taken from his mother's Tuscan hamlet near Vinci to grow up in his father's house in Florence. When Ser Pierro's third wife bore Leonardo's father a second, but legitimate, son, Leonardo was automatically disinherited from any claim on the family fortune and was on his own.

As a child born out of wedlock, Leonardo would be denied the right to enter a profession, obtain a position at the university, or gain social access to the royal courts. On top of this, Leonardo was determinedly left-handed in an era when people were taught that left-handedness was literally sinister — a tool of the devil — and something to be avoided. Leonardo dealt with this by perversely writing from right to left, and backwards. Was this to avoid smudging the writing or to keep prying eyes from readily reading his words? Or was it Leonardo's way to draw attention to his ideas, his sketches — and himself? And if illegitimacy and left-handedness were not enough to cope with, Leonardo had to confront a third challenge in the years to come: He was almost certainly homosexual at a time when such behavior was subject to severe penalties and public ostracism.

Given Leonardo's status and prospects in society, his father hoped that the boy's great interest and obvious talent for drawing might qualify him for an apprenticeship. It did. Popular artisan Andrea del Verrocchio headed one of Florence's leading craft studios — where everything from furniture to religious objects, from portraits to statues — was crafted for the city's aristocracy and wealthy merchants. His apprenticeship to one of Florence's leading artists would give him the training and experience that could serve him well in the future — as well as help him rise above the circumstances of his birth.

Leonardo served as Verrocchio's assistant for nearly nine years; during that time he learned a number of different crafts. Leonardo also came to understand the role of risk and failure in making progress. He welcomed failure as his way of finding a path to ultimate success. This attitude set the stage for the development of the scientific method that remains the guiding principle for so much of our progress today.

He was given the honor of providing an angel in the corner of a painting that became known as the *Baptism of Christ*. Legend has it that Leonardo's contribution to that painting was so extraordinary in execution that Verrocchio himself pledged never to paint anything else ever again.

In the atmosphere of Verrocchio's all-male studio, Leonardo's left-handedness may well have been embraced for the authority it gave his brush strokes, and his physical beauty may well have been welcomed by some of his colleagues. Because of his acceptance in the studio and its relative safety from the harsher attitudes of the outside world, Leonardo did not hesitate to take great risks in his work and in his thinking. These intellectual, artistic, and personal risks clearly provided satisfying rewards that may well have been the stimulus for the great advances that he would make later and that we marvel at today.

For example, in the field of art he was one of the first to experiment with oil-based paints because of their flexibility and luminosity. He mastered the crucial skill of perspective to create detailed background scenes and he experimented with blending light and dark areas seamlessly to give his paintings a more realistic look. He invented a surface for frescoes that permitted him to work at his own pace rather than within a time frame dictated by traditional methods. In his desire for attention and recognition, he also showed an unbridled willingness to do whatever was asked of him to gain the approval of his patrons and their friends. We never think of him as such, but in modern terms his desire to please helped Leonardo become a superb event planner, a skilled costume designer, an acclaimed lighting technician, a resourceful impresario, and an accomplished musician.

When Milan was faced with the threat of invasion by the French in the late 1490s, Leonardo volunteered to develop machines and engineering strategies to assist Duke Ludovico Sforza to repel the foe. Many of the machines Da Vinci sketched and described — for war or for engineering purposes — were probably never built because they required too much money, too many scarce materiels, or too many experiments to perfect their use or performance.

Perhaps because of his insatiable curiosity, perhaps because he could never seem to harness his mind to concentrate on one topic until he completely mastered it, perhaps because of his overpowering need for perfection, he kept looking for ways to improve the performance of all manner of machines. His patience and persistence revealed a pattern that would mark all of Leonardo's efforts as a designer and engineer: If it wasn't perfect, it wasn't finished. Because of this, he rarely saw projects to completion. In fact, the younger Michelangelo criticized Da Vinci for procrastination. Was this seeming indolence triggered by his perfectionism or by his constant inquisitiveness? Or might it have been a part of his survival instincts borne of his outcast status — as long as something was incomplete, he would be needed to finish it.

Anything that he observed in nature could trigger a flood of questions in his mind about how it functioned, how it accomplished particular tasks, and the purpose served by each of its attributes. None more so than his fascination with birds to unlock the secrets of flight. His desire to be free from the confines of earth suggests the confidence he had that there was nothing he couldn't accomplish if he put his mind to finding a solution. For Leonardo, nature was the perfect machine that had to be understood completely in order to be emulated as closely as possible in man-made efforts. That is why he became skilled as an anatomist — to learn how the muscles, tendons, and skeletal structure worked together to make a smile or raise an arm, why he theorized on how the fossils he found high in the Alps might have gotten there, and why the few paintings he brought to completion — only 14 are known to have survived to the modern era — were renowned then as now for the subtly, accuracy, and secrets they embodied.

The best way to form your own impression of this remarkable man and what he accomplished is to remember what he once proclaimed:

The natural desire of good men is knowledge.

and to contemplate what others have said about him. Consider the memorable words of author E. L. Kongsburg's :

[Leonardo] could not look at things made by God
without wondering how He had made them,
and he could not look at things made by man
without thinking of some way to make them better.

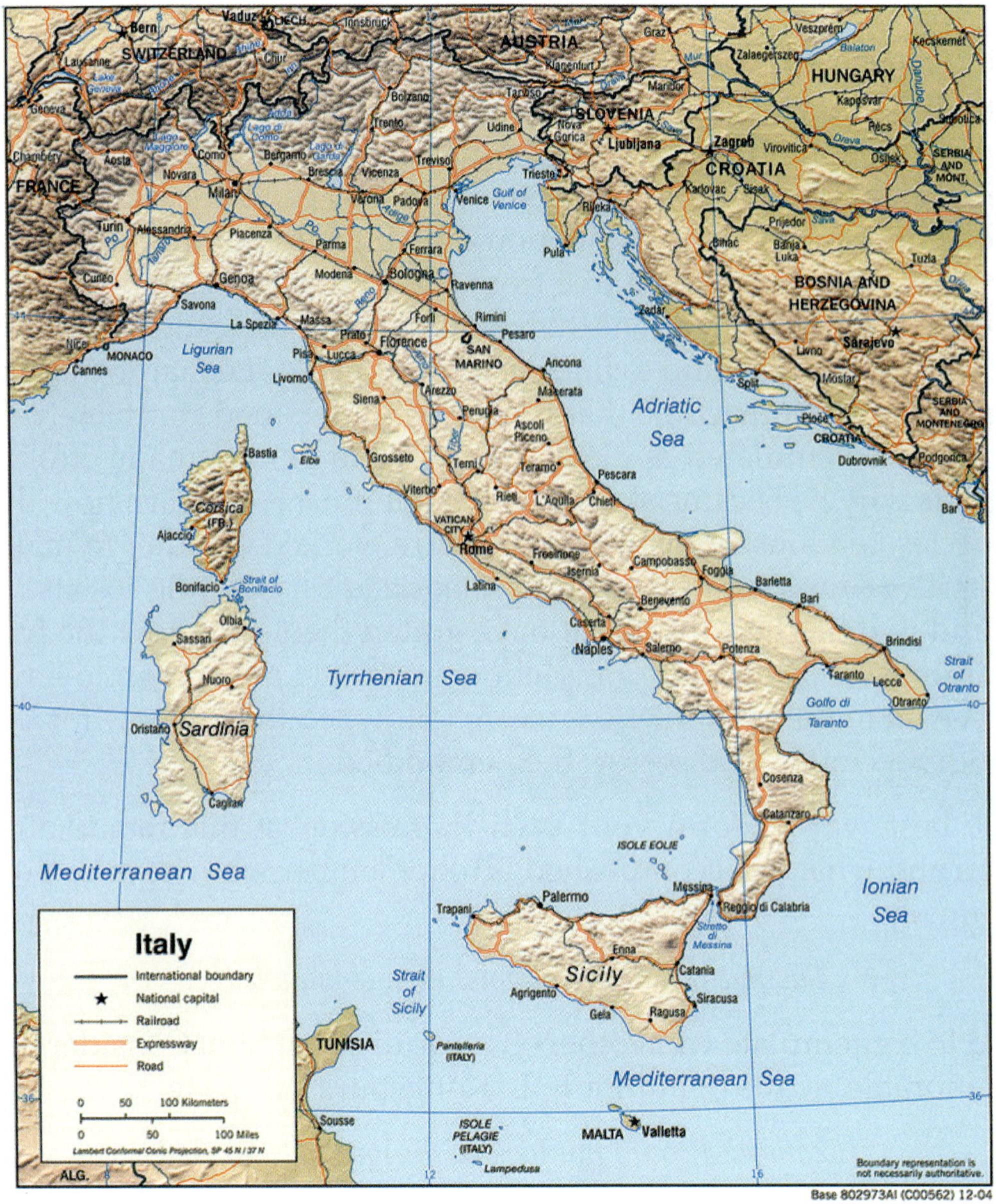

Modern Italy

TIMELINE OF LEONARDO DA VINCI'S LIFE

YEAR	LEONARDO'S LIFE	ELSEWHERE
1452	Born out-of-wedlock at a villa in the hamlet of Anciano. Barred from taking his father's name, he becomes Leonardo "of Vinci."	
1453		Constantinople and the Byzantine Empire fall to the Turks, and many intellectuals and their scholarly manuscripts find their way to Italy.
1458		Guttenberg introduces movable type for Western printing presses.
1467	Begins a 12-year apprenticeship in Andrea del Verrocchio's Florence studio.	
1472	Accepted into Guild of Florentine Painters and starts *The Annunciation*—his first major painting.	
1475	Ser Piero, Leonardo's father, has a legitimate son by his third wife, ending Leonardo's hope for any inheritance.	

YEAR	LEONARDO'S LIFE	ELSEWHERE
1475		The rifle is invented.
1476	Contributes angel early on to Verrocchio's *Baptism of Christ.* Accused of sodomy, the charges were eventually dropped for lack of evidence.	
		Inca Empire gains control of most of South America.
1477		The Netherlands, Luxembourg, and Burgundy join the Holy Roman Empire, based in Vienna.
1478	Leonardo receives first commission as an independent artist at the age of 26. Paints *Ginevra de' Benci.*	
1481		Mongols defeated and driven from Russia.
1482	Leaves Lorenzo d'Medici's Florence for Milan and becomes an employee of Duke Ludovico Sforza. Leonardo plays the lyre and remains some 18 years as	

YEAR	LEONARDO'S LIFE	ELSEWHERE
	an artist, event planner, and engineer in residence. Most of Leonardo's labor-saving machines and mechanical devices were designed during this period of his life.	
1484	Begins his serious study of flight.	The plague sweeps Europe, killing huge numbers of people.
1485	Paints *Lady with an Ermine*, a portrait of Cecilia Gallerani, thought to be the Duke's mistress. Involved with painting *Virgin of the Rocks*.	Bartolomeu Dias sails around the Cape of Good Hope on behalf of Portugal. The Wars of the Roses between two houses of the English throne ends.
1487	Begins his daring study of anatomy with his first autopsies.	
1489	Draws the *Vitruvian Man* exploring harmonious proportions.	

YEAR	LEONARDO'S LIFE	ELSEWHERE
1492	Leonardo visits Rome for the first time.	Columbus reaches the New World.
1493	Builds clay model for huge bronze horse as tribute to Duke Francisco Sforza, Ludovico's father; it is unveiled at the wedding celebration of Sforza's neice.	
1495	Leonardo begins work on *The Last Supper* in a Milanese monastery; he finishes this work in 1497.	The Ming Dynasty, having revolutionalized agriculture and built the Great Wall, starts to lose power in China to merchants and traders.
1496	Ready for his first flight.	
1498		The French capture Milan.
1499	French soldiers destroy the clay horse, using it for target practice after Sforza's palace is captured.	
	Leonardo presents his ideas for underwater defenses to the	

YEAR	LEONARDO'S LIFE	ELSEWHERE
	Duke of Venice to protect against a Turkish naval invasion. The ideas are rejected.	
		Vasco da Gama sails around Africa to reach India.
1500	Rivalry with Michelangelo becomes public. Begins *Virgin and Child and St. Anne.*	
		The Half-Millennium recognized and celebrated by the Catholic Church.
1502	Becomes military engineer to Cesare Borgia, Duke of Florence, and Pope Alexander VI's son. Borgia is the model for Niccolò Machiavelli's *The Prince.*	
		The first small portable clock is invented.
1503	Begins the *Mona Lisa,* works on it for the rest of his life. Commissioned to paint the *Battle of Anghiari.*	
		Columbus stranded in Jamica on final voyage to the New World in search of a Western passage to India.

YEAR	LEONARDO'S LIFE	ELSEWHERE
1504	Leonardo meets Michelangelo; later in the year, Leonardo's father dies.	
1507	Francesco Melzi, Leonardo's faithful assistant, begins working for the master. Leonardo describes the effects of arterio-sclerosis for the first time.	
		King Henry VIII accedes to English throne.
1509	Leonardo draws detailed maps and conducts his geological studies.	
		France declares war on Venice, but the Pope soon falls out with Louis XII and allies himself with Venice to drive the French out of Italy.
1510	Complete *Virgin and Child and* St.Anne.	Coffee, as a social drink, reaches Mecca.
1513	Leonardo returns to Rome.	
		Michelangelo completes Sistine Chapel ceiling at Vatican.

YEAR	LEONARDO'S LIFE	ELSEWHERE
1513		The Holy Roman Emperor honors the memory of the Dutch fisherman who first pickled herring—an important method of preserving food at the time.
1515	Leonardo's mechanical lion is presented to the King of France as a gift from the Duke of Urbino and the following year moves to Chateau Amboise in France to work for King Francis I until his death.	
1517	Leonardo suffers a stroke paralyzing his *right* side.	
		Martin Luther calls for reform of the Catholic Church.
1519	Leonardo dies on May 2 in France.	
		Hernando Cortez arrives in Mexico and begins conquest of the Aztecs.

THE DA VINCI EXPERIENCE EXHIBIT

BACKGROUND ON THE DA VINCI EXPERIENCE EXHIBIT

When we were first offered the opportunity to manage a display of replicas of the machines designed and described by Leonardo da Vinci—built and organized by artisans in Florence—we decided to present the exhibit in terms of the context of this renowned man's life. It was our belief that Leonardo was not just a painter, not just an engineer, not just a scientist in the service of royalty in the 15th century, but an ordinary man who accomplished extraordinary things in great part because he was driven to overcome personal doubts and family uncertainties.

He did so, we believe, because he was willing to take enormous risks throughout his career—in his attempts to find a way to fly, in the autopsies he performed, in the way he decided to paint religious icons, in his determination to improve on machines in common use, and in the ideas he developed and discussed. We thought that all of this offered an important but different role model for young people. How Leonardo overcame the difficulties inherent in his birth and upbringing fascinated us. It matched a study we had published of the first 100 years of Nobel Prize winners, more than 65 percent of whom also experienced challenging childhoods similar to Leonardo's. In overcoming their difficulties, they too matured faster and produced more than their contemporaries.

Because he dared convention, Leonardo also came to understand the role risk-taking played in making progress. Risk led to failure, but each failure taught a lesson for the next attempt to get closer to success. Leonardo's attitude toward risk and failure seemed to us in sharp contrast to modern approaches. How many young people are sheltered today from risk by parents who fear failure might harm a child's self-esteem? Yet, sheltered young people in Japan commit suicide at the highest rate in the world—some 32,000 a year; some Western kids tend to deal with their first failures and disappointments through alcohol and drugs. Leonardo, on the

---

* Shalev, Baruch, *100+ Years of Nobel Prizes and More,* The Americas Group, 2010

other hand, learned to welcome failure as a way to ultimate success.

There was another startling contrast between Leonardo and modern attitudes. Modern scientific personalities tend to be honored for their expertise, which often results in an extraordinarily narrow focus. Leonardo, on the other hand, wanted to diversify his knowledge as broadly as possible.

We knew from the start of our involvement in this project that a museum delivers learning in a vastly different way from such other educational forums as classrooms, books, or briefings. Museums allow individuals to explore and learn from the displays at their own pace and in their own way. Combine this open, unstructured approach with a display heavily weighted toward interactivity between visitor and artifact and you have an unparalleled opportunity to achieve a greater appreciation and deeper understanding of the information you wish to impart.

We were then faced with the question of where we might find a museum to host the Da Vinci Experience. After all, the major metropolises were filled with facilities offering a wide variety of permanent and temporary displays of fine art or specialized objects, all of which are continually competing with dozens of other leisure activities for the attention of urban citizens. We decided that if we could site our collection of Da Vinci machines in communities within the orbit of a major metropolitan area, we would have the nearly undivided attention of local residents, the full support of that community's tourism industry, and interested attendees coming from larger metropolitan areas. If the exhibit proved worthy of the effort it would take to see it, those who made the commitment to attend would become part of the oldest, least expensive, most powerful marketing force available; they would become the source of the word of mouth comments we knew we would need to build an audience of sufficient size to meet the expectations of our suppliers and investors.

With that audience in mind, we decided to display the 60 machines selected for our exhibit in three galleries: transportation, military, and mechanical. In that way, we could disburse the interactive devices to prevent crowding. We also hoped to provide visitors with displays that would allow them to savor the breadth of Leonardo's interests, the linkages between what we now have and what Leonardo was trying to provide 500 years ago, and the relationship he saw between the pursuit of art and the development of science.

Those thoughts led us to another consideration. How could one present ideas and objects conceived so many years ago with the wonder that they deserve today? After all, Leonardo had no electricity, no plastics, no computers to help him realize his dreams. We decided the answer was to introduce visitors to the display by bringing them out of their modern air-conditioned environment into a re-creation of a 15th century Florentine passageway. It didn't work. People hurried past our effort to see the introductory film on Leonardo's life and work. We also decided to put the displays illustrating Leonardo's other interests at the *end* of the displays by leading visitors into a reproduction of a Florentine Palazzo's hallway where reproductions of Leonardo's paintings would be on display. Because so much of the exhibit was designed to excite kids, we created and have continually erxpanded an Educational Area with a wide variety of puzzles, games, and "make and take" projects. Visitors can also enjoy a special gift store filled with a wide variety of books, apparel, souvenirs, toys, and other items devoted to the many aspects of Leonardo's legacy.

There are many people to thank for their contributions to making the Da Vinci Experience exhibit possible. While all of them are listed in the Acknowledgments at the back, we would draw special attention to Kenneth Katz, the co-manager of Da Vinci Experience, LLC, and a true friend, experienced businessman, and always reliable associate; Michael Mathews, our sponsor consultant and gift store manager; and Luigi Rizzo, a colleague in various enterprises, who brought us the Da Vinci exhibit from his work with Gabriele Niccolai and the Museum of Leonardo da Vinci in Florence.

Los Angeles
January 2009

Godfrey Harris
Curator

CATALOG NOTES

The following notes on how the exhibit has been compiled and organized may prove helpful in understanding this catalog:

- The names of the machines used in the exhibit and in this catalog are somewhat arbitrary. We have tried to be as descriptive of the purpose and workings of each machine as possible. Leonardo did not spend a lot of time on nomenclature. He drew and sketched ideas without bothering to label many of them. The Italian names for the machines follow the identifications and numbering system used at the Museum of Leonardo da Vinci, where the replicas were first on display. The English names used in New Zealand, Australia, and elsewhere may be somewhat different from the names used here.
- Leonardo drew multiple sketches and offered many versions of his machines over a number of years. The replicas pictured in this catalog and seen in the Da Vinci Experience exhibit are from sketches or notes selected by the artisans at the Museum of Leonardo da Vinci in Florence. In some cases, the artisans have made several versions, each faithful to a description and sketch provided by Leonardo. Hence, a replica seen in this catalog may not match a replica on display in another museum or at another exhibit.
- Leonardo captured his thoughts and ideas in numerous sketchbooks — on what is believed to be more than 14,000 pages. After Leonardo's death in 1519, Francesco Melzi, his assistant, carefully cataloged and preserved these notebooks. But after Melzi died, individual elements of the sketchbooks were disbursed. They were subsequently brought back in multiple new volumes known as codices—sometimes in an organized fashion, sometimes in a jumble of unconnected subjects. Leonardo didn't make subsequent organization easy. He often mixed multiple subjects on both

sides of a single sheet of paper. Modern scholars have given these known codices names with slight variations to accommodate for English, French, Spanish and Italian usage. This catalog has adopted the following names for its references:

- Codex Atlantico
- Codex Trivultianus
- Codex Marciano
- Codex on Birds
- Codex Arundel
- Codex Foster
- Windsor Collection
- Madrid Manuscripts
- Paris Manuscripts
- Codex Leicester

These are located in various repositories in Europe and the United States. The first two above — Atlantico and Trivultianius — are in Milan; Marciano is in Venice. Codices Arundel, Foster and Windsor are in the United Kingdom. Codex Leicester is in Seattle, the property of Bill Gates, the Chairman of Microsoft. Individual Da Vinci sketches that were never put into one of the codices are known to be in both private hands as well as museum collections.

Many of the sketches of the machines in the Da Vinci Experience were discovered in the Madrid Manuscripts. Because it is impossible to date the individual Da Vinci notes or drawings, most scholars rely on dates given to the various codices themselves. For example, the items in Codex Trivultianus were created circa 1487 to 1490; the items in the Madrid Manuscripts, circa 1509, were gathered during Leonardo's French period.

- The sequence of codex names, numbers and letters to identify the location of a particular Da Vinci sketch has been developed by scholars and the owners of the codices. The references here are the same Italian language versions utilized by the Museum of Leonardo da Vinci in Florence. For example, the machine entitled "Flight" was derived from drawings and / or notes referenced as *Codex Atlantico f. 824v.* The name of the codex (*Codex Atlantico*) is followed by a folio or page reference (*f*) and the number of the folio (824). The final letter in the sequence refers to the position on the page where the sketch or note is to be found: "v" is a Roman numeral and means that it is the 5th drawing on that folio or page; the letter "r" stands for "retro" or the back of the folio or page; an asterisk — * — means that the drawing or note is found on the front of the folio or page. Some references to a Codice refer to a letter with the word "Manoscritto," such as Manoscritto B. These refer to the Paris Manuscripts which are lettered from A to M. When no Roman numeral appears in the numbering sequence, the drawing is the only one on the page. Readers should note that Codex Atlantico was recently renumbered and this may result in some discrepancies between books using the old classification system and others using the new one.

- Each machine also carries a number. "Flight" is numbered LV 19. This number is given to each replica by the Museum of Leonardo da Vinci and refers to the order in which the machine was studied and built by the artisans of Niccolai S.N.C. for the Museum of Leonardo da Vinci.

THE DA VINCI EXPERIENCE

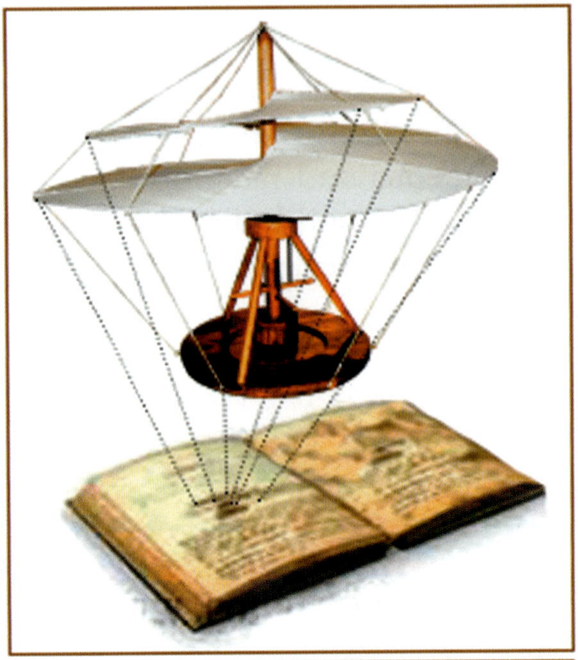

Machines Designed and Described by
LEONARDO da VINCI

THE MACHINES OF THE TRANSPORTATION GALLERY

AIR SCREW

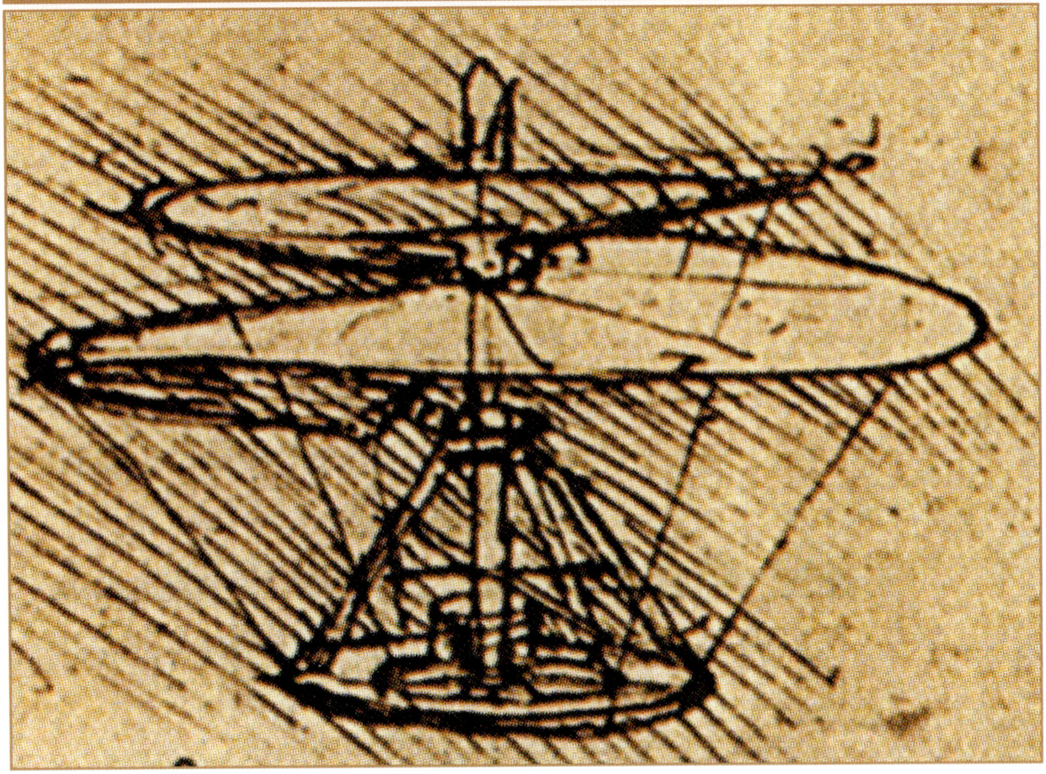

AIR SCREW
(Vite aerea)
LV 21
Manoscritto B, f. 83 v

AIR SCREW

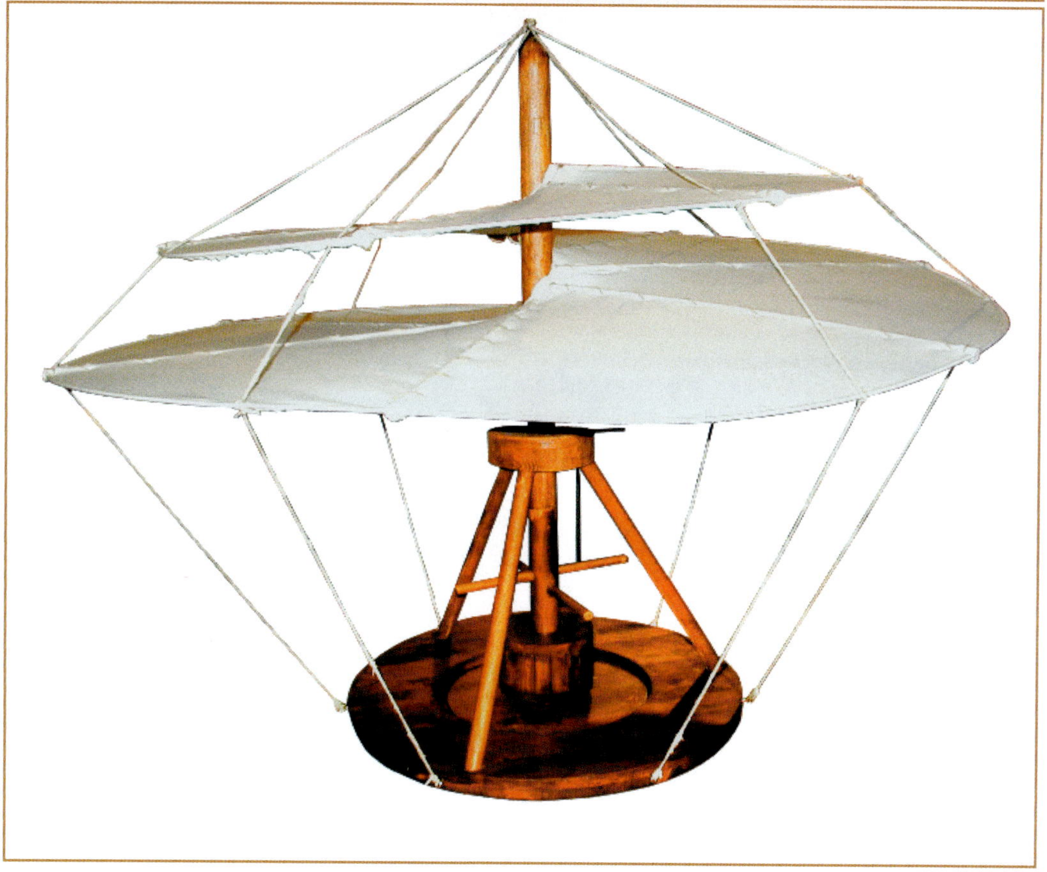

Air Screw: Leonardo's air screw design is a forerunner for the modern helicopter — a machine that lifts off from the ground and rises straight up. Leonardo thought air currents could keep the craft aloft once he generated enough energy to get it started. Da Vinci had the right idea, but men would need another four hundred years to learn how to produce the amount of energy actually needed for the task.

ANEMOMETER

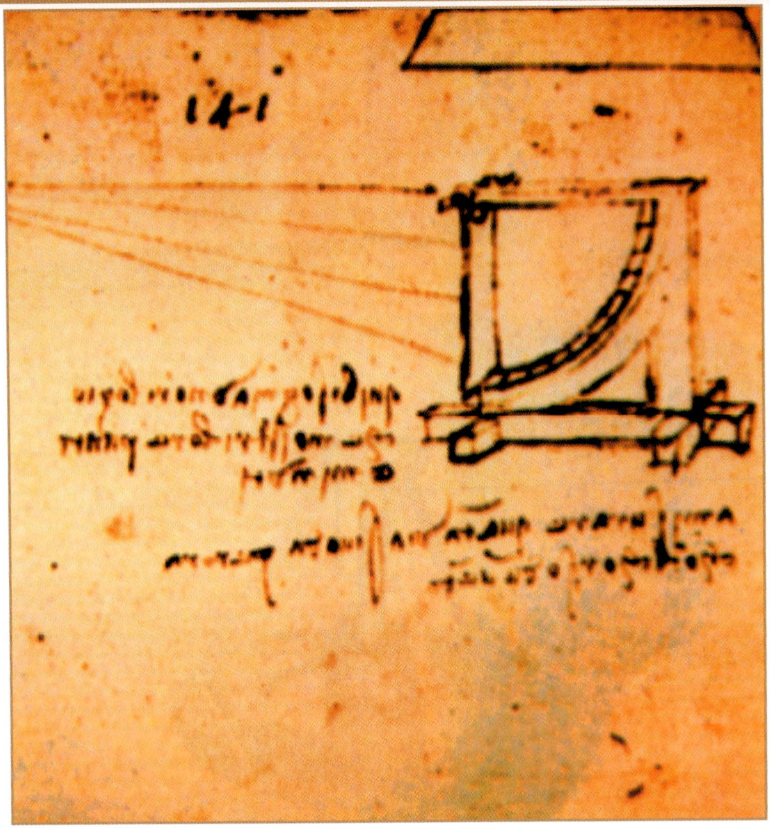

ANEMOMETER
(Anemometro a Pennello)
LV 03/04
Codex Atlantico f. 675

ANEMOMETER

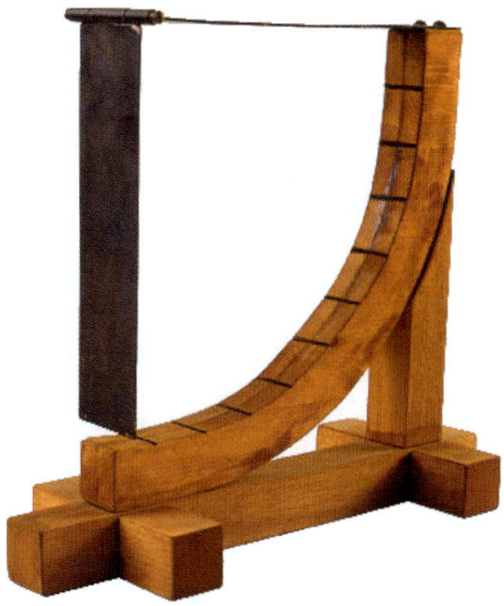

Anemometer: This is an instrument for measuring the wind's speed. As the wind strikes the steel plate, pushing it back on its hinge, the speed can be calculated on the scale carved into the stand behind.

BICYCLE

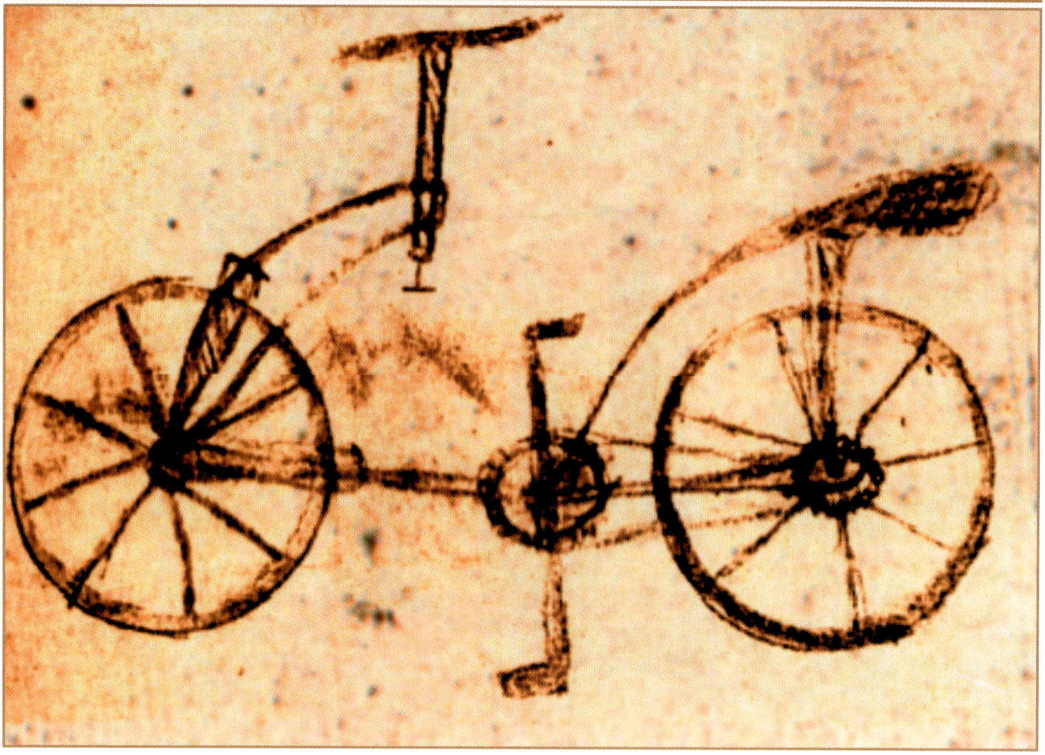

BICYCLE
(Bicicletta)
LV 31
Codex Atlantico f. 133 v

BICYCLE

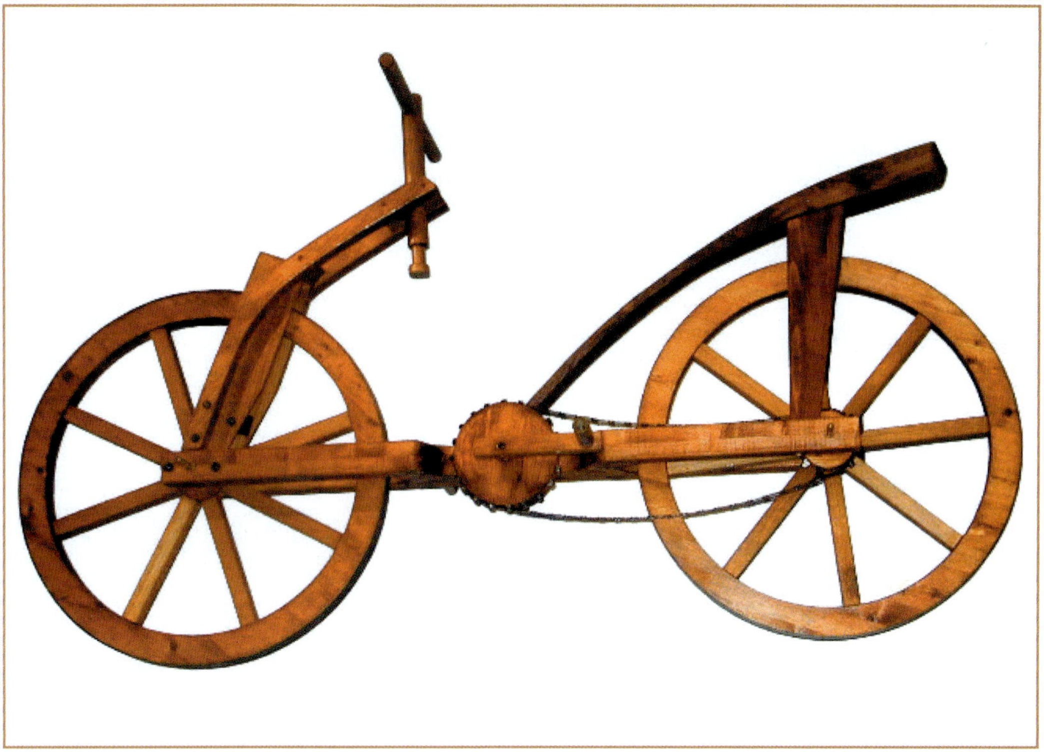

The Bicycle: This is one of the many mysteries still surrounding Leonardo's life and work. During restoration of the Codex Atlantico, the drawing of what we now call a bicycle was found between two glued-in pages. The lack of detail in the drawing of the bicycle and the signature of one of Leonardo's pupils on it lead some to think that the drawing is not by Leonardo himself but by a student who may have copied the drawing from a model seen in the workshop.

CAR

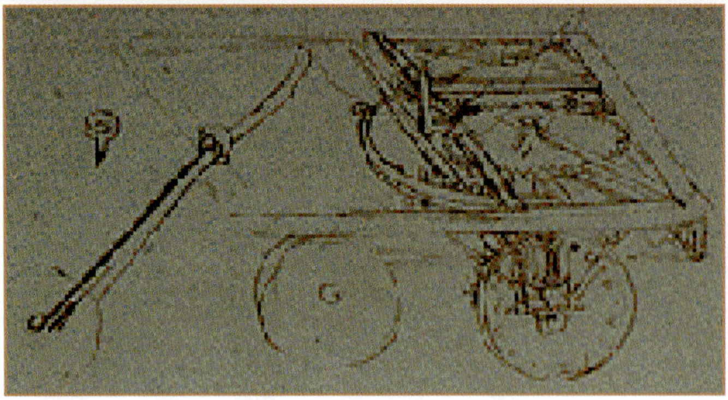

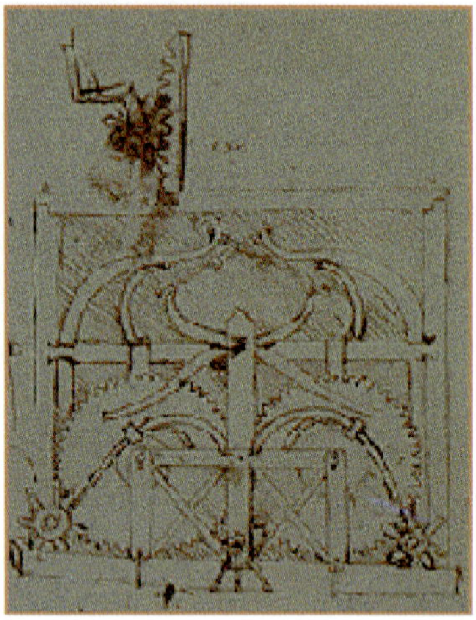

CAR
(Macchina a balestre)
Codex Atlantico f. 812 r

CAR

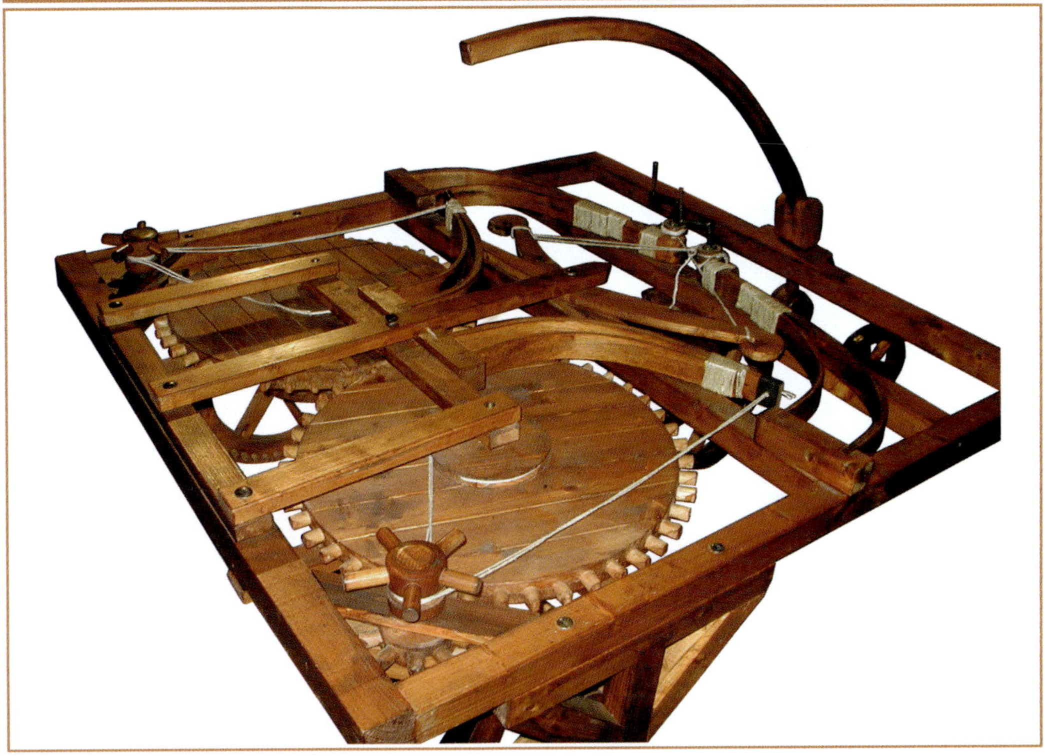

Car: This is one of the most famous of Leonardo's inventions. It was designed to be powered by leaf-springs (balestre) and programmed to move in a specific pattern. While the car was self-propelled, an operator would have been needed to tighten the leaf springs to power the wheels. The stored energy in the leaf springs is transmitted to the driving wheels by means of a complex set of gears. A small single wheel at the rear is controlled by a rudder to provide a means of steering the car.

CHAIN

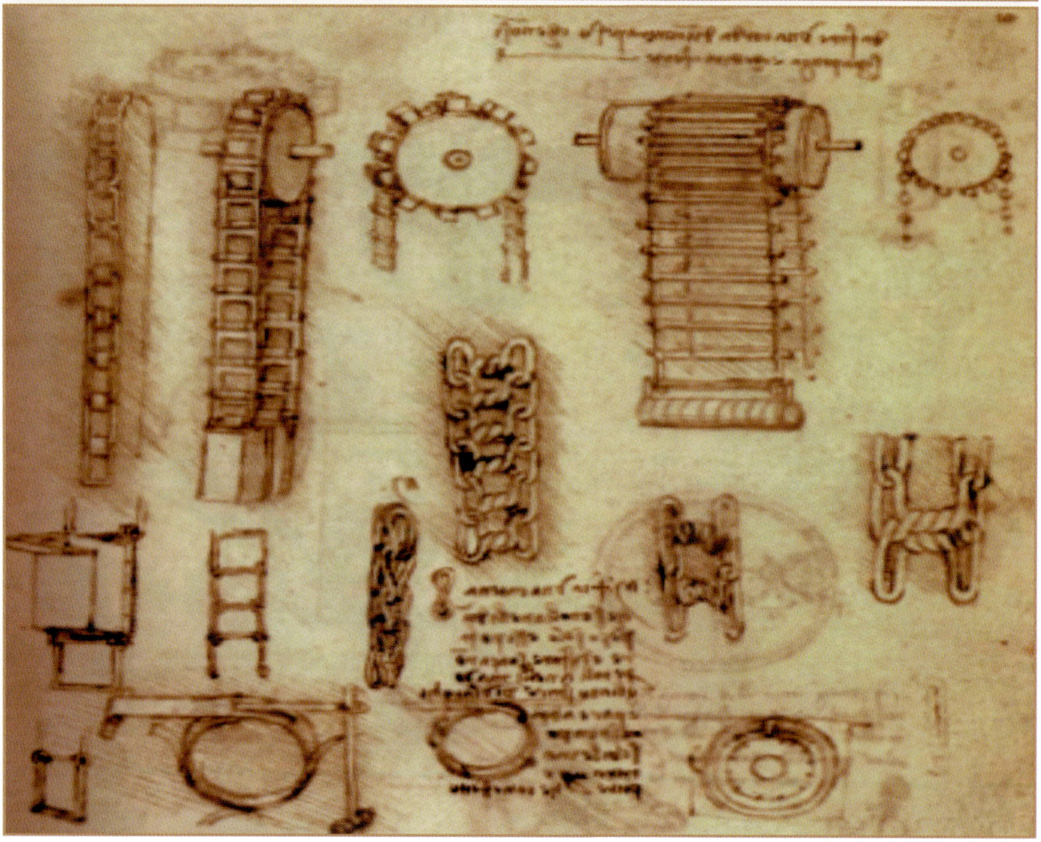

CHAIN
(Catene)
LV 09
Codex Madrid I f. 10 r

CHAIN

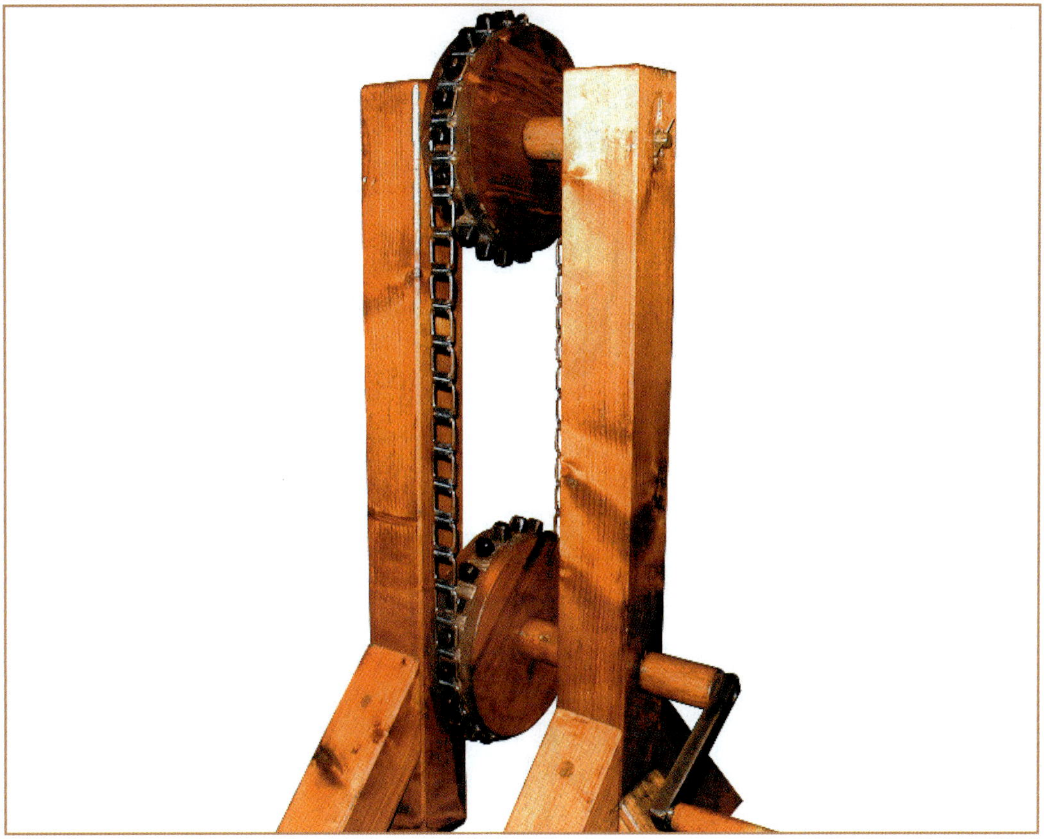

Chain: Leonardo studied many different types of flexible chains as a means of transmitting continuous motion from one point of a machine to another. The chain is a key feature of one of Da Vinci's most famous machines: the bicycle.

DIFFERENTIAL

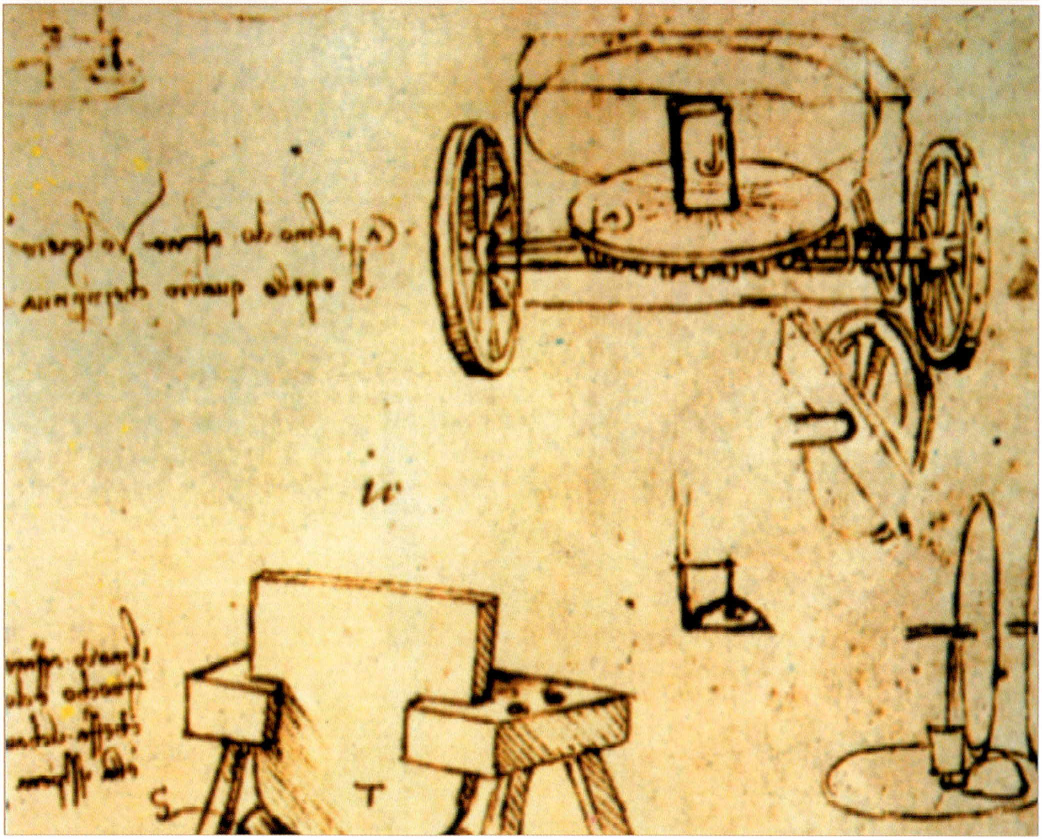

DIFFERENTIAL
(Carro con differenziale)
LV 49
Codex Atlantico f. 1049

DIFFERENTIAL

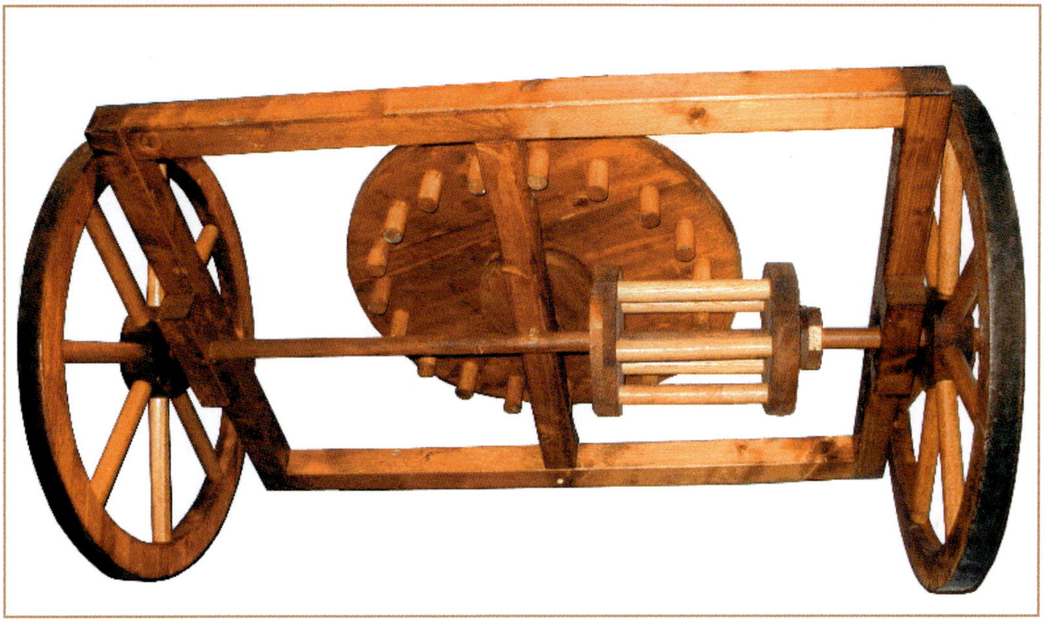

Differential: Leonardo understood a fundamental problem of all two-wheeled carts. In curves and turns, the outer wheel has to go further and move faster than the inner wheel. To solve the problem, Leonardo devised a way to transmit motion through the axle connecting the two wheels. A crank turns the cog wheel which engages the lantern rocket linked to a single wheel, thus increasing its velocity. In its modern version, as illustrated below by Jamie Pfeifer Graphics, the same function is performed by gears attached to the drive shaft and rear axle.

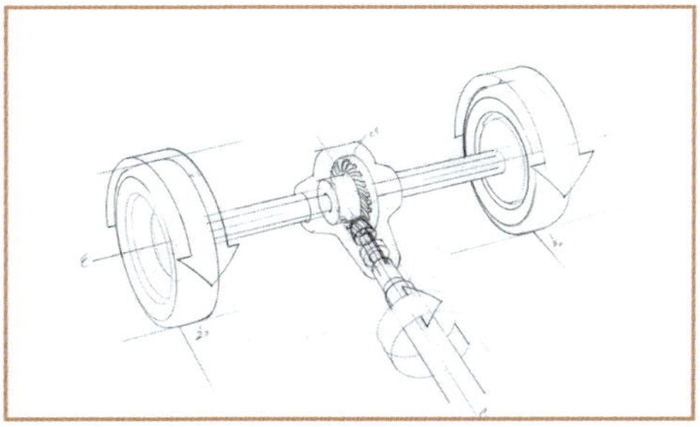

DOUBLE-HULLED BOAT

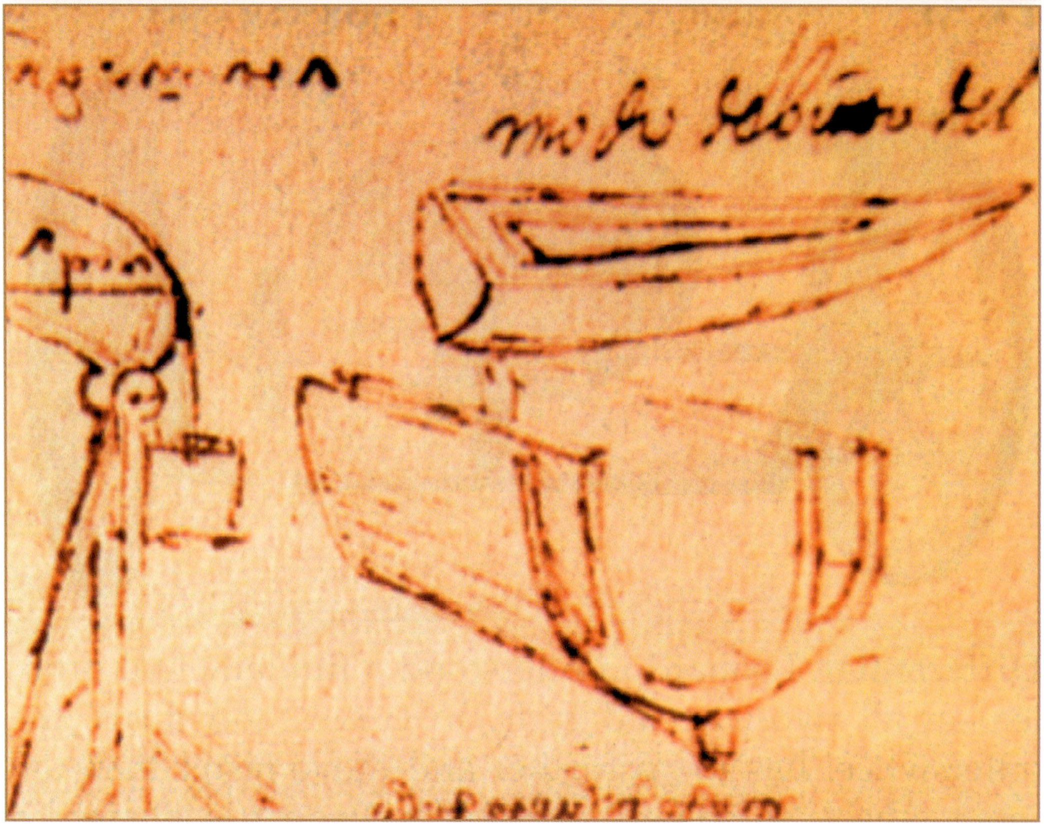

DOUBLE-HULLED BOAT
(Imbarcazione a Doppio Scafo)
LV 24
Manoscritto B f. 11 r

DOUBLE-HULLED BOAT

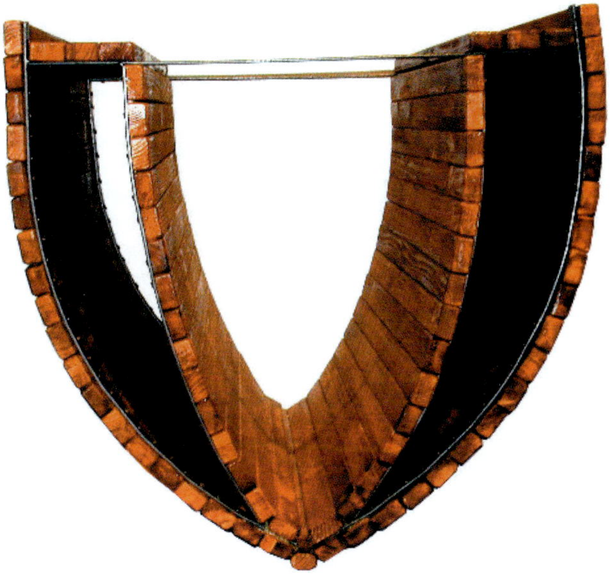

Double-Hulled Boat: While double-hulled construction is now a standard design feature for tankers carrying bulk liquids, Leonardo saw the concept as a defense mechanism to protect warships during battles at sea. Perhaps no other machine designed and described by Leonardo da Vinci shows both the originality of his imagination and the simplicity of his ideas as clearly as this boat design.

FLAPPING WING

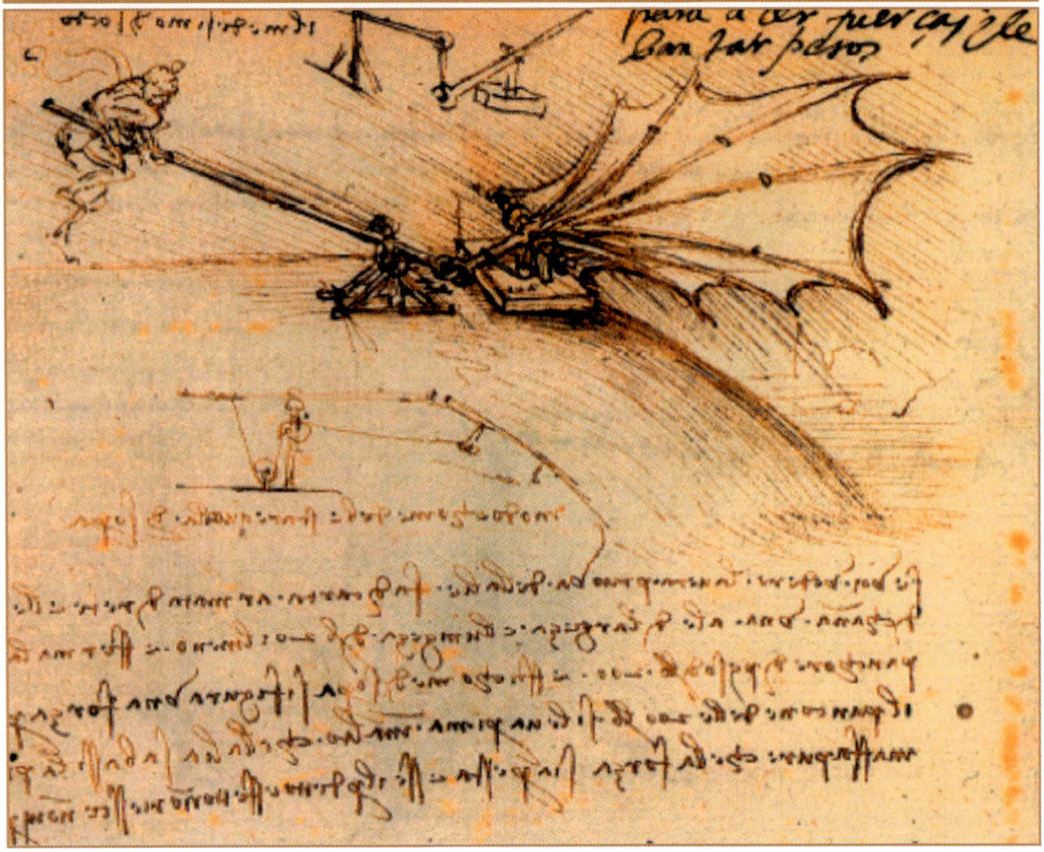

FLAPPING WING

(Ala Battente)

LV 20

Manoscritto B f. 88 v

FLAPPING WING

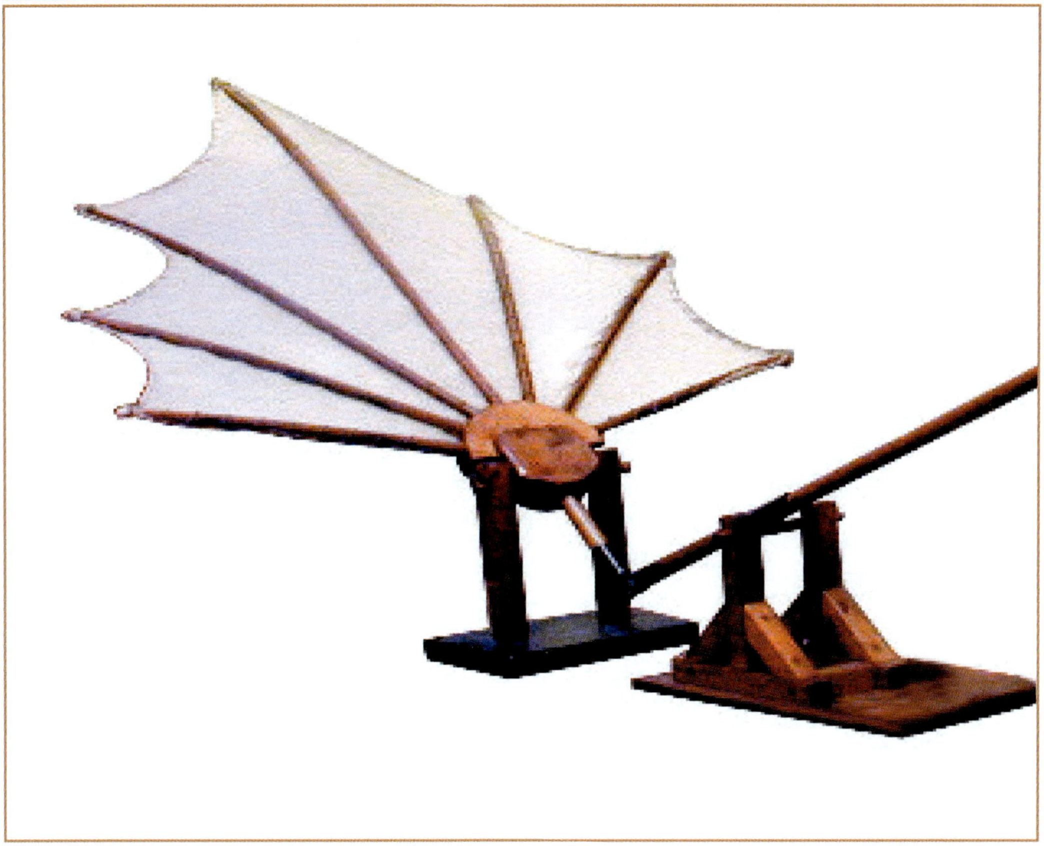

Flapping Wing: This machine depicts one of Leonardo's most intriguing experiments. He thought that if the long lever could be lowered rapidly enough, the wing would be able to lift the plank — weighing about the same as a man. Leonardo calculated the ratio of the wing span to the weight of the plank on the basis of his observations of birds. As a result, the wing would have had to be a massive 12 meters long by 12 meters wide — about the dimensions of a large 40 foot room — to do the task he envisaged for it.

FLIGHT

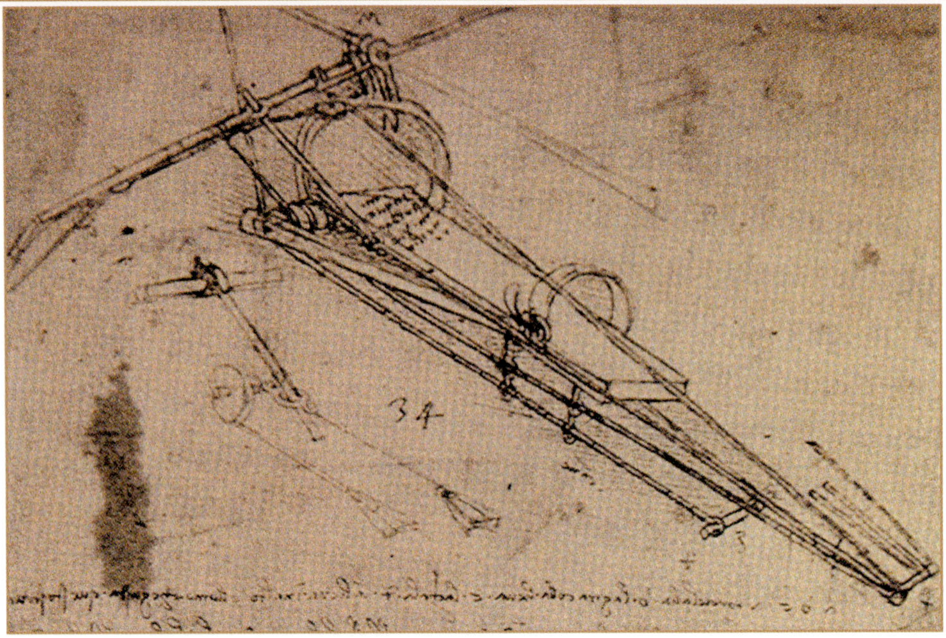

FLIGHT
(Macchina Volante)
LV 19
Codex Atlantico f. 824 v

FLIGHT

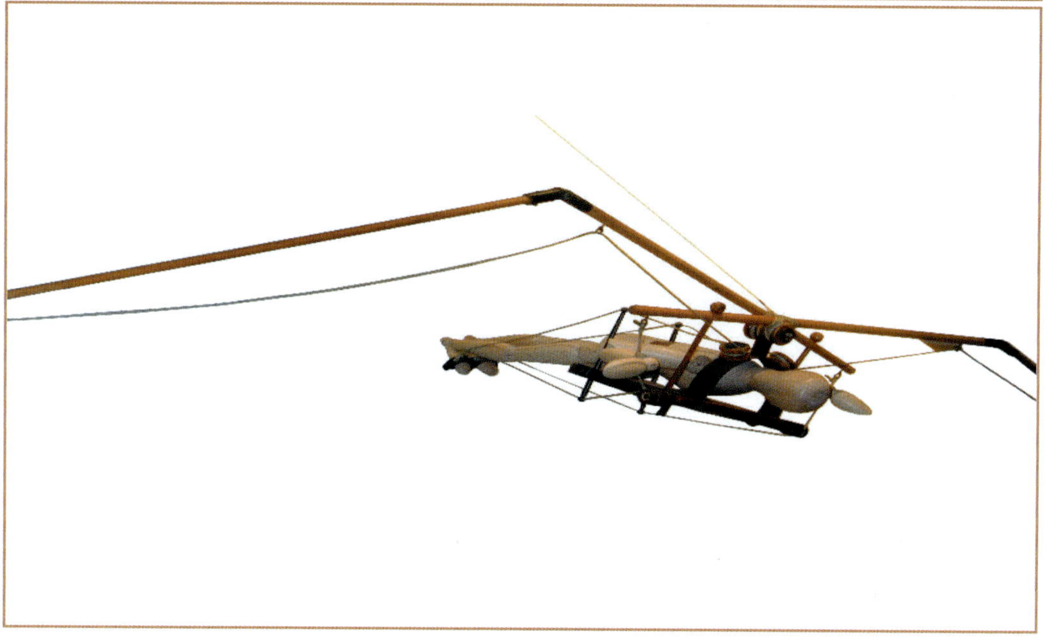

Flight: This is one of Leonardo da Vinci's most famous creations. The machine's pilot lies stretched out on a platform under the supporting wing. With his feet in the stirrups, the pilot is able to raise and lower each wing to guide flight.

GEAR SHIFT

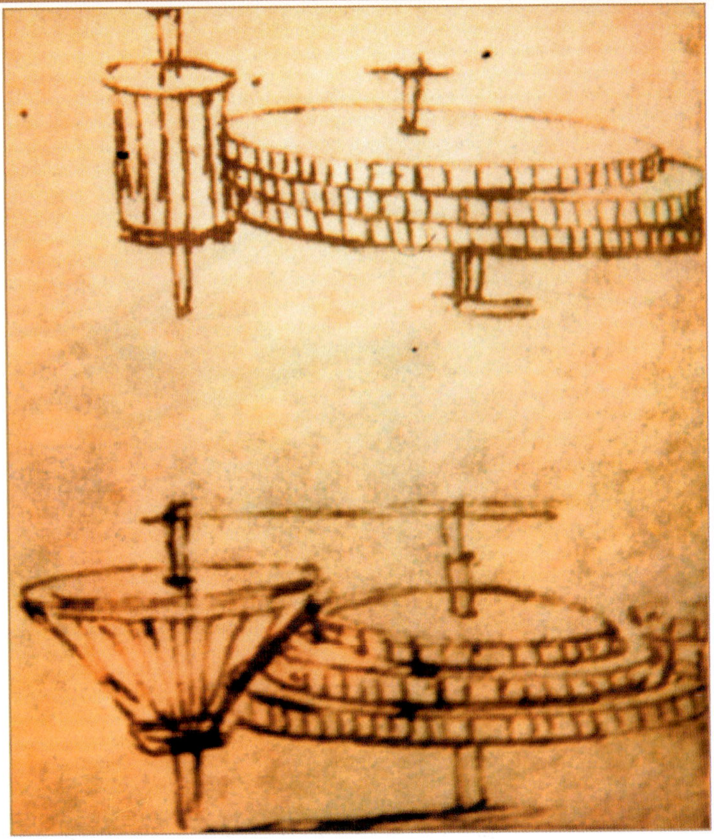

GEAR SHIFT
(Cambio di Velocitá)
LV 46
Codex Atlantico f. 77 v

GEAR SHIFT

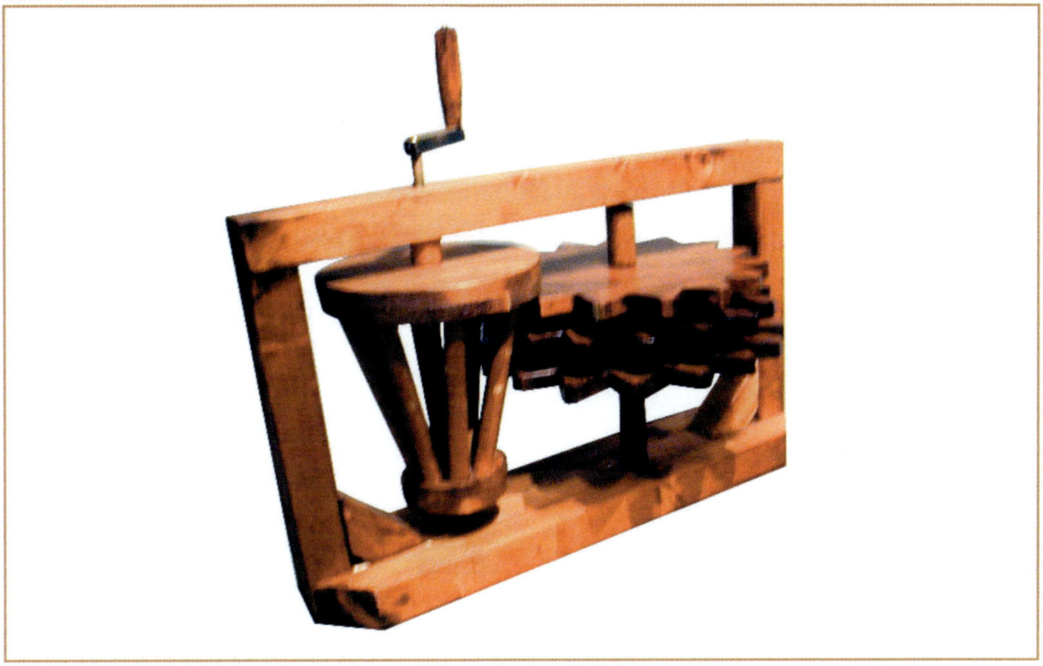

Gear Shift: This gear was intended to focus high energy to achieve a speed that could be maintained with less energy. In this example, a conical cage (lantern wheel) transmits motion to cog wheels with different diameters. Although their rotational velocity to complete a full revolution is the same, each wheel generates a different speed. This is the same principle used in transmissions of modern cars (See illustration below.)

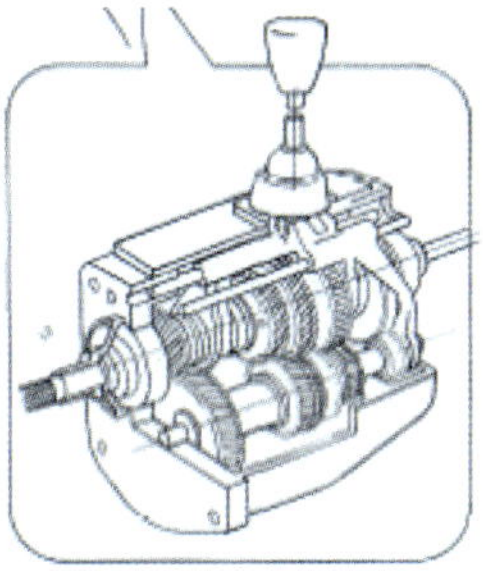

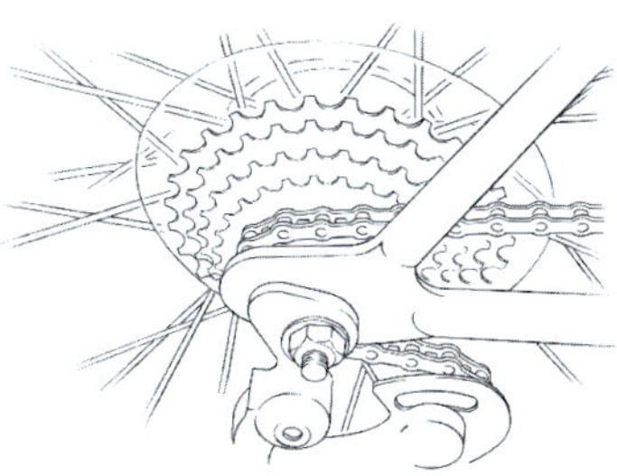

Illustrations by Jamie Pfeifer Graphics

A view of a modern automobile transmission (l.) and a three-speed bicycle gear (r.) shows how they are based on the same principle as Leonardo was exploring.

GLIDER

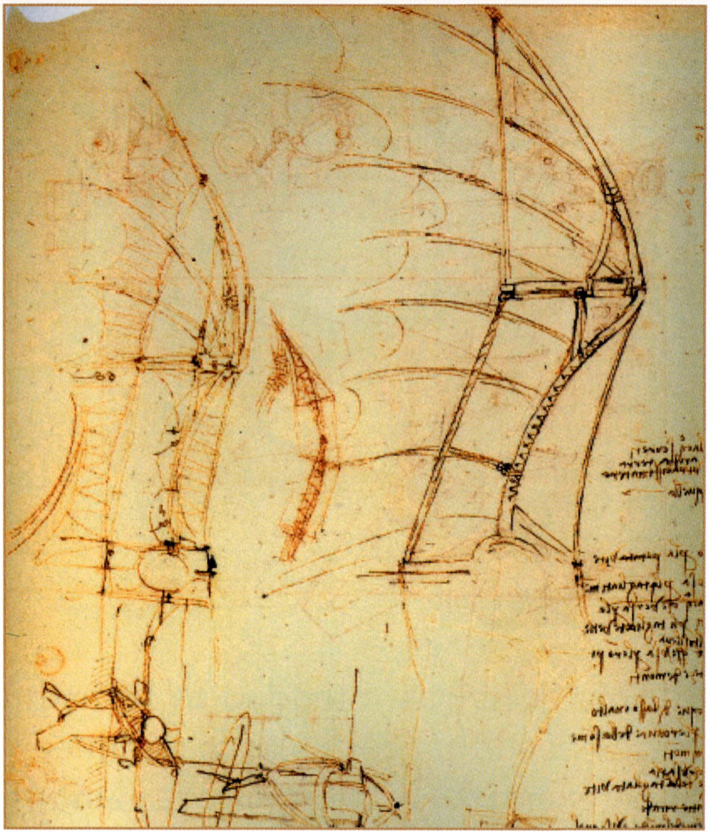

GLIDER
(Aliante)
LV 23
Codex Atlantico, f 846 v

GLIDER

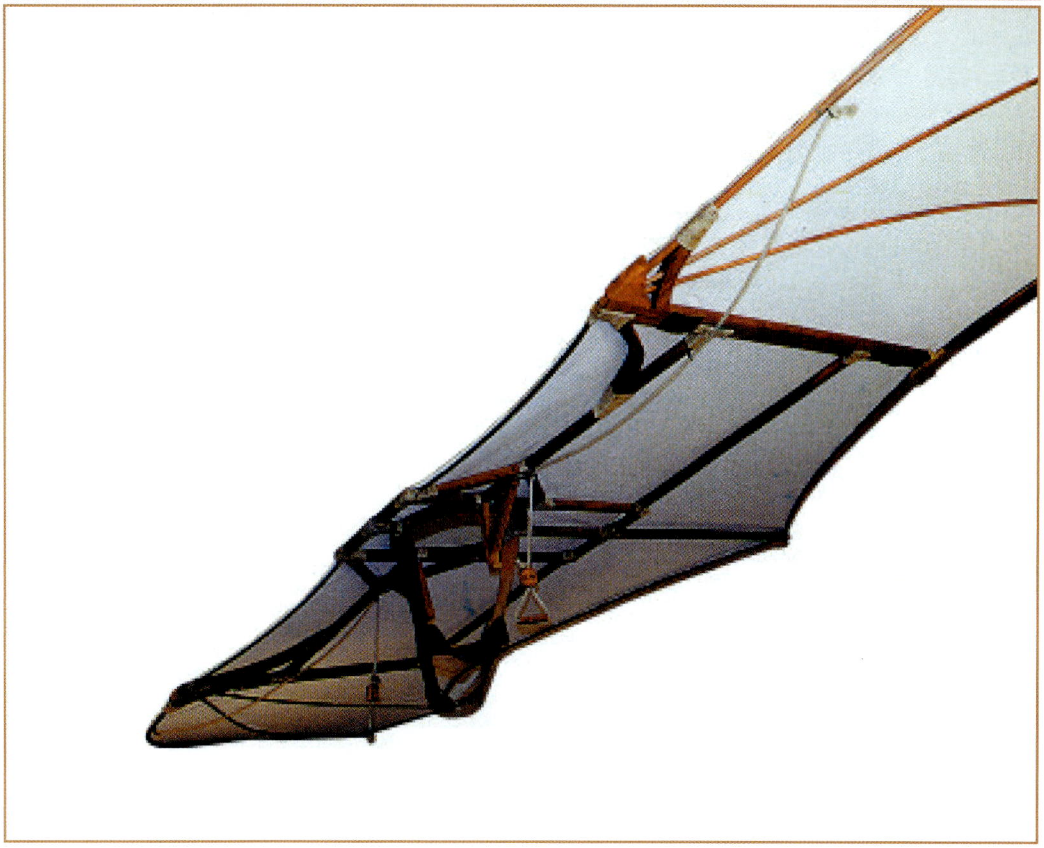

Glider: In this craft, the pilot assumes a balanced position in the center. The wings are fixed at the center while the extremities are operated by cables manipulated by the handles. After observing the flight of birds, Leonardo understood that the inner part of the wing moves more slowly than the outer part, requiring the inner part to provide support for the outer.

HANG GLIDER

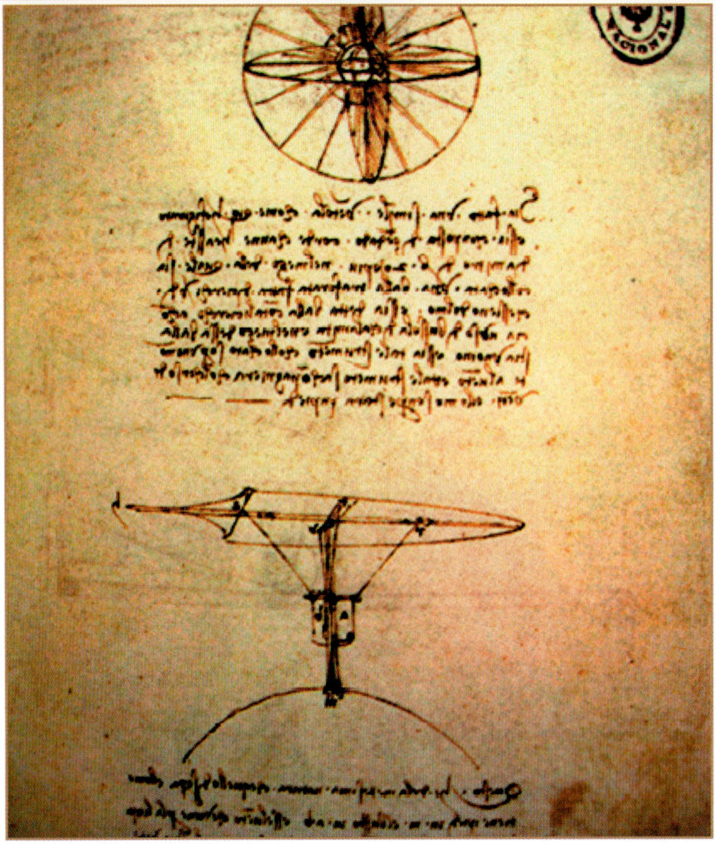

HANG GLIDER
(Deltaplano)
LV 48
Codex Madrid I f. 64 r

HANG GLIDER

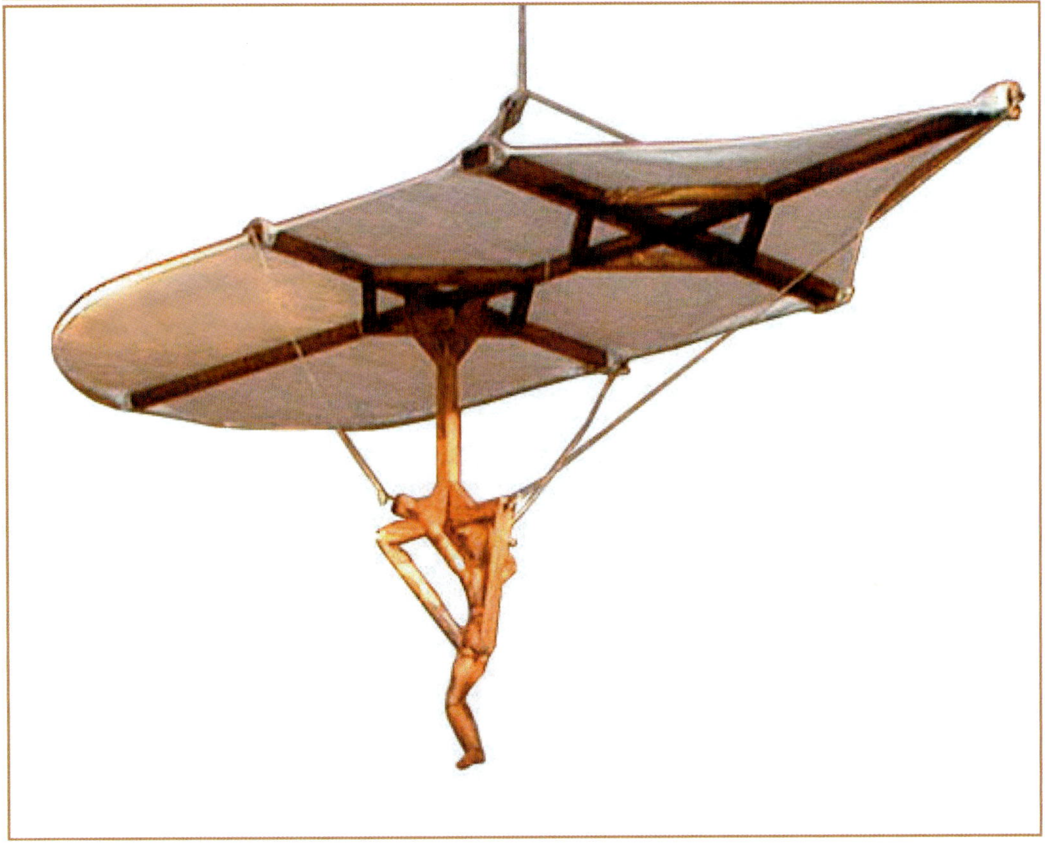

Hang Glider: Leonardo studied flying machines in two different periods of his career. In the first period, his flying machines were designed to fly on the basis of the strength of a man. The machines that belong to the second period utilized wind power. The hang glider belongs to the second period. Note the rudder-like tail that permits the pilot to regulate and control the flight while soaring.

HELICOPTER

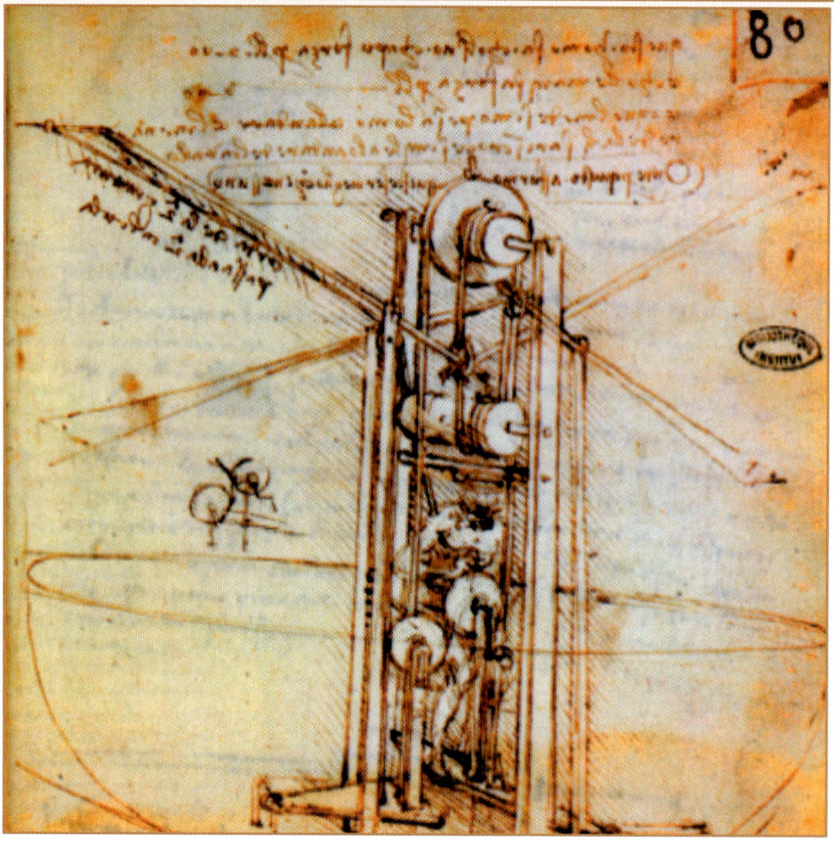

HELICOPTER
(Ornitottero Verticale)
LV 64
Manoscritto B f. 80 r.

HELICOPTER

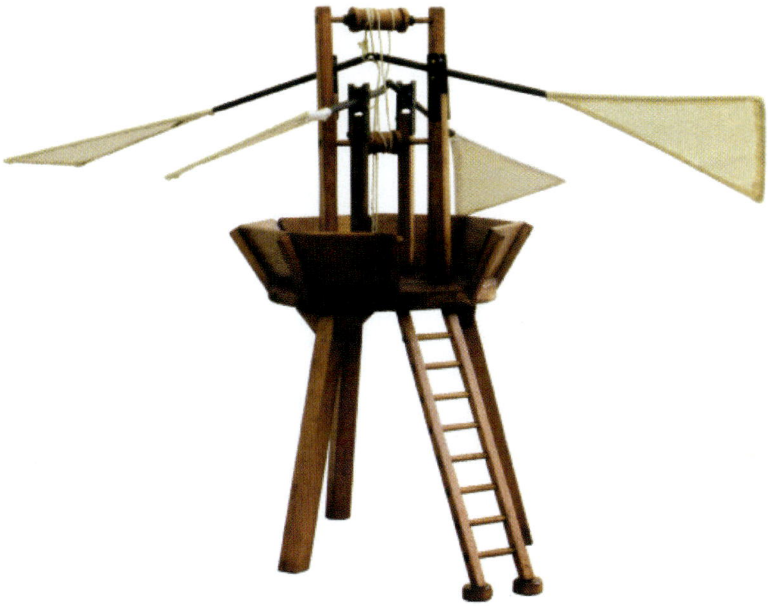

Helicopter: In this concept, Leonardo envisaged a pilot standing upright in the airship. To control the machine during flight, the pilot was to use his hands, feet and head. Due to the size and weight of the device with a pilot aboard, Leonardo decided to utilize two crossing pairs of what we now think of as rotor blades.

INCLINOMETER

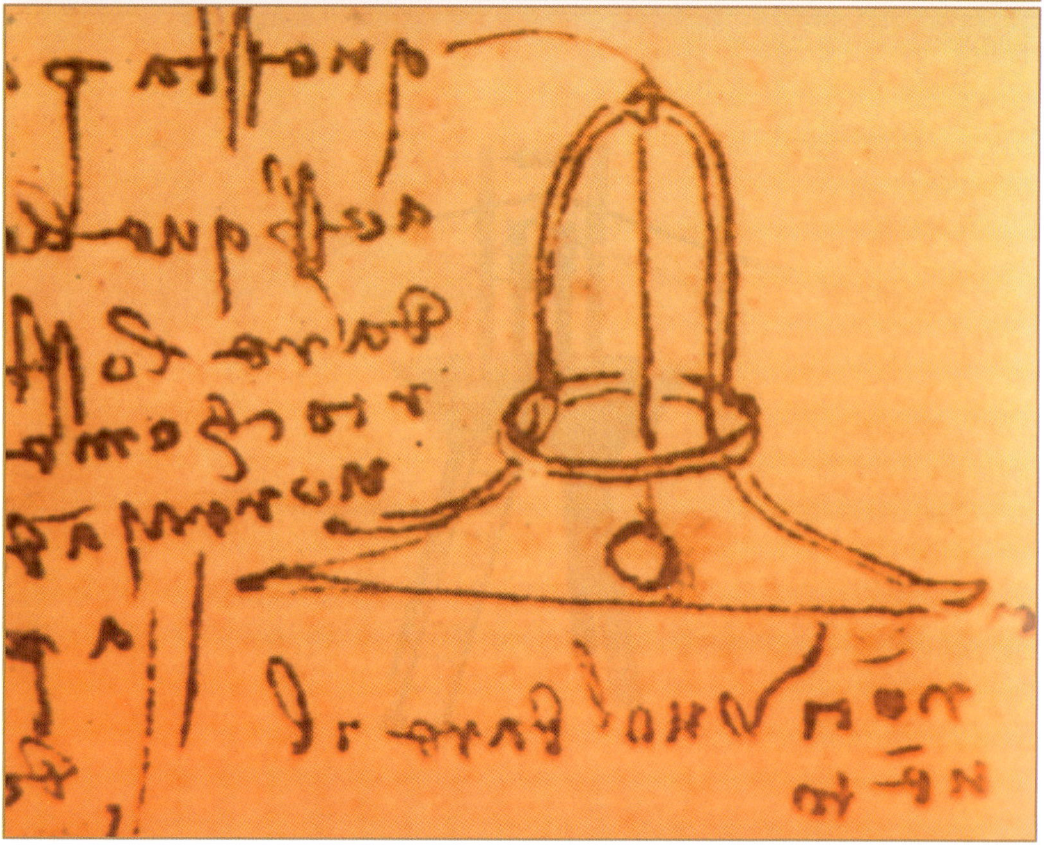

INCLINOMETER
(Inclinometro)
LV 05
Codex Atlantico f. 1058 r

INCLOMETER

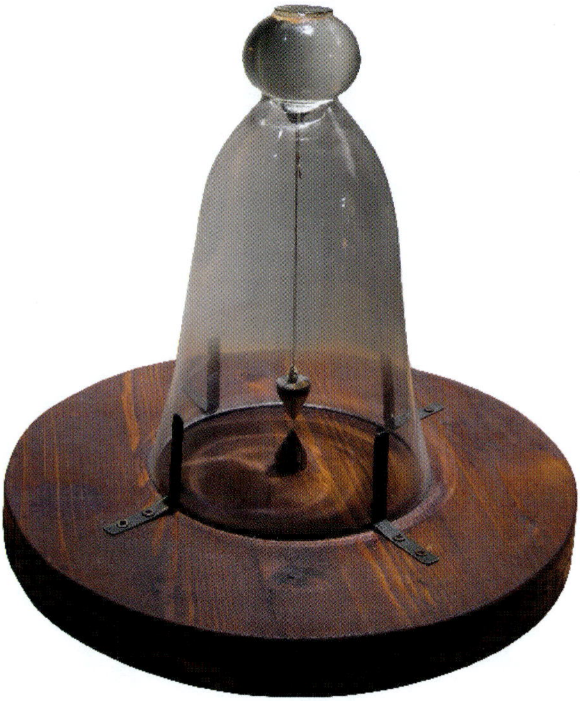

Inclinometer: This is an instrument whose purpose is still in modern use: to show the physical position of an airplane or ship relative to a perfect horizontal orientation. As the platform on which the weight rested dipped front to back or side to side, the weight would come to rest at an observable angle. The bell jar protected the weight from the effects of air currents.

JACK

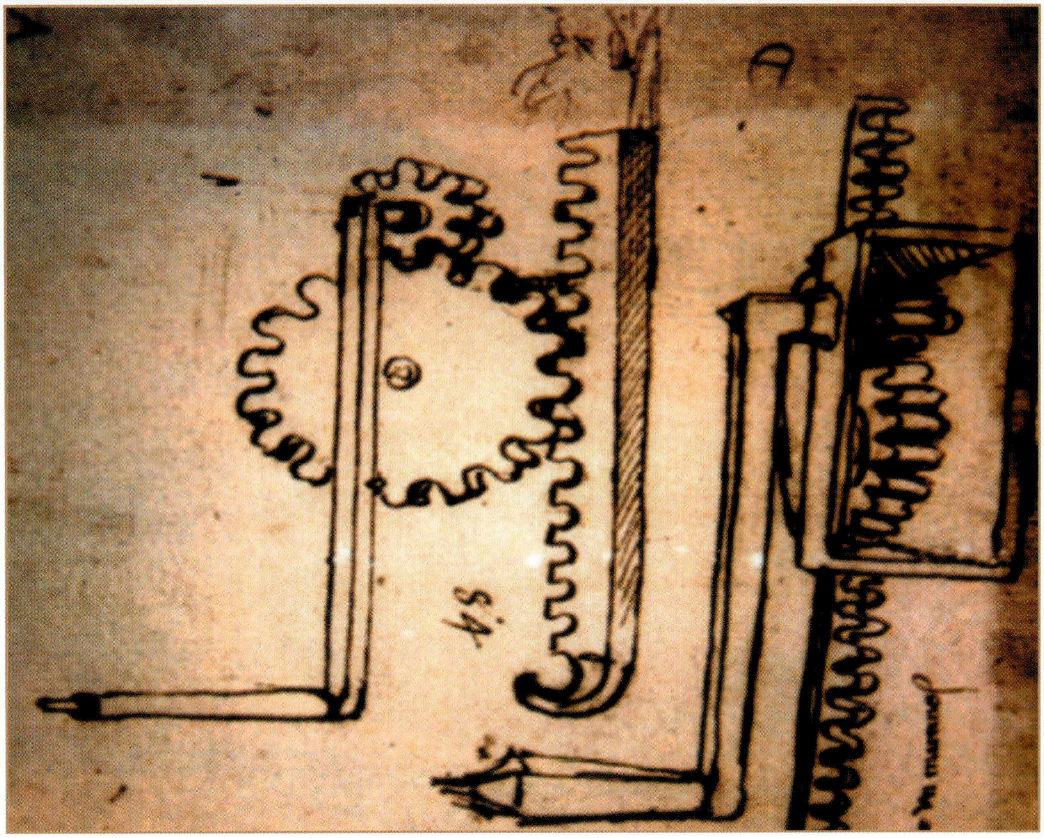

JACK
(Cric)
LV 59
Codex Atlantico f. 998 r

JACK

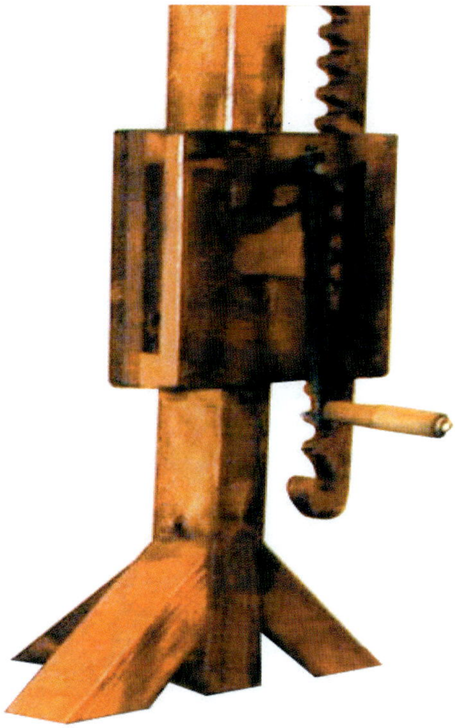

Jack: This machine was conceived by Leonardo to lift heavy objects. Anyone who has had to change an automobile tire understands the principle. A crank linked to a reduction gear works a rod with teeth that moves up or down to raise or lower the attached load.

LIFE PRESERVER

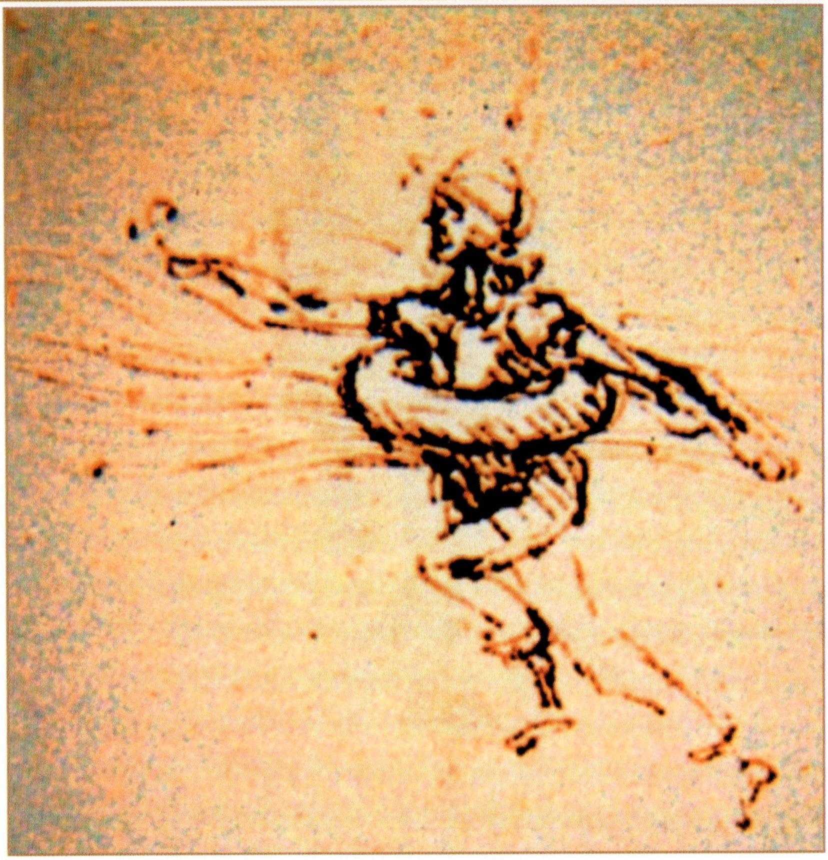

LIFE PRESERVER
(Salvagente)
LV 54
Manoscritto B, f. 87 v

LIFE PRESERVER

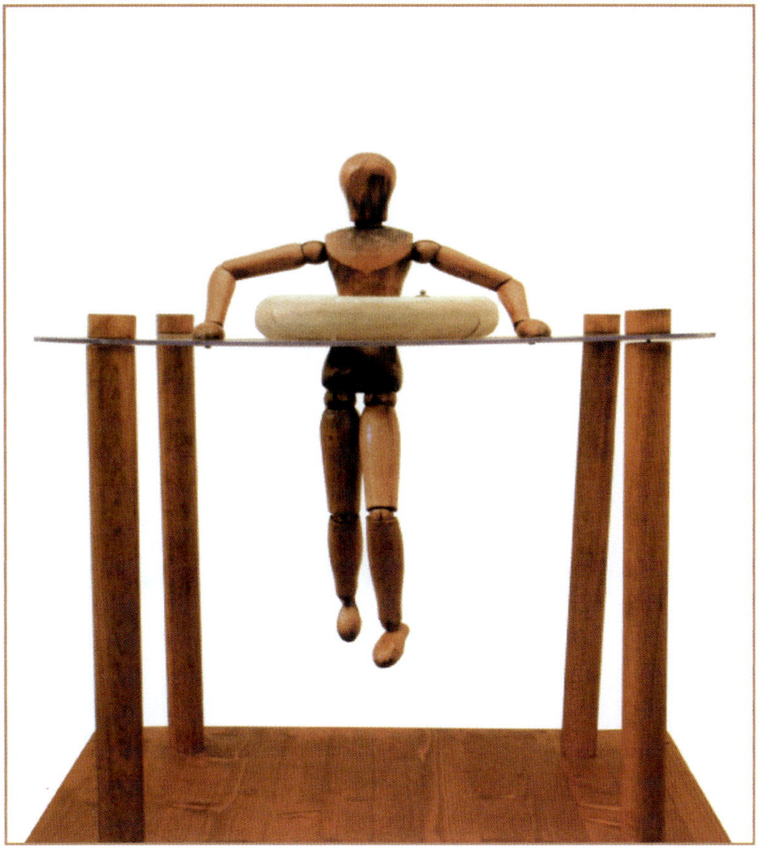

Life Preserver: This is one of Leonardo's many ideas to enable man to work and persevere in a water environment. His life preserver was made of waterproof leather and could be filled with air to increase its volume and buoyancy.

PADDLE-WHEELED BOAT

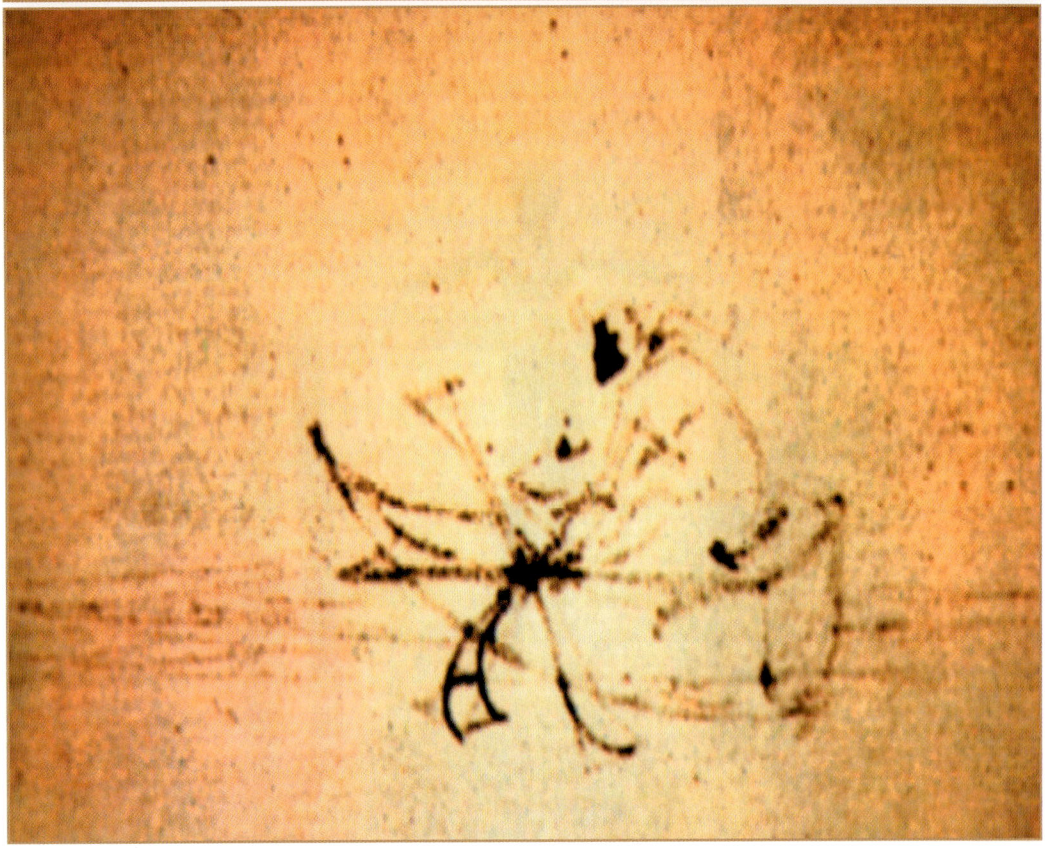

PADDLE-WHEELED BOAT
(Barca a Pale)
LV 25
Manoscritto B, f.83 r

PADDLE-WHEELED BOAT

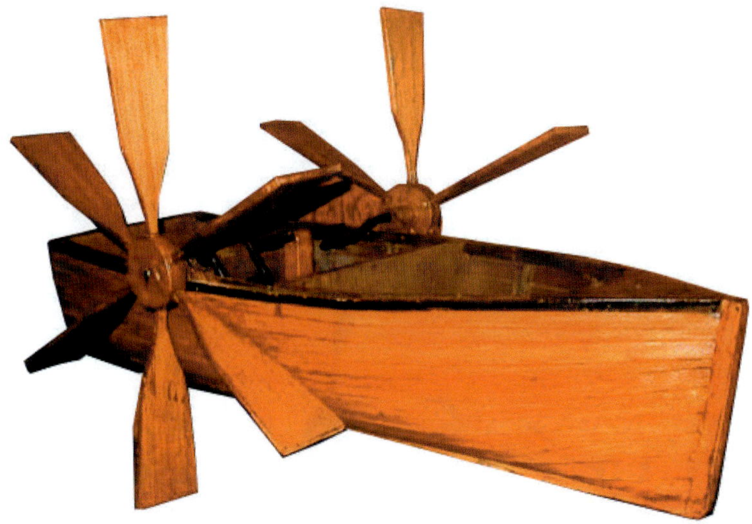

Paddle-Wheeled Boat: Leonardo sought to increase the speed and navigational precision of boats. The shape of the hull was of primary importance in this issue, and Leonardo used the shape of a fish to guide him. He also thought of equipping some boats with large paddles that could be worked by hand or foot cranks, perhaps aided by flywheels, thus improving the rhythm and efficiency of traditional oars. In the concept presented here, the prow is tapered and maneuverability is enhanced because the crank operators face forward.

Leonardo's use of a crank to transmit power can also be found in the pedal mechanism of a modern exercise bicycle.

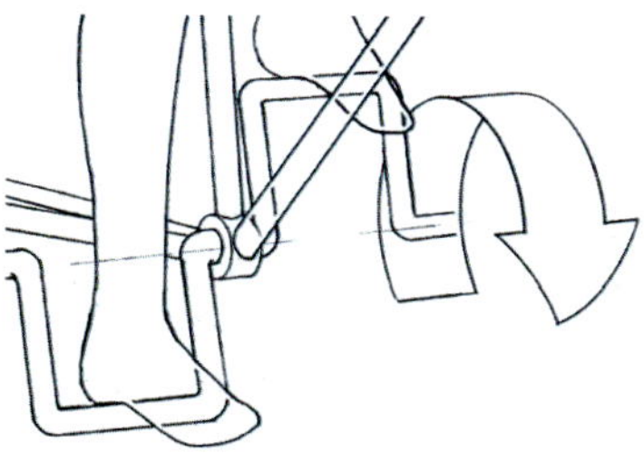

Illustration by Jamie Pfeifer Graphics

PARACHUTE

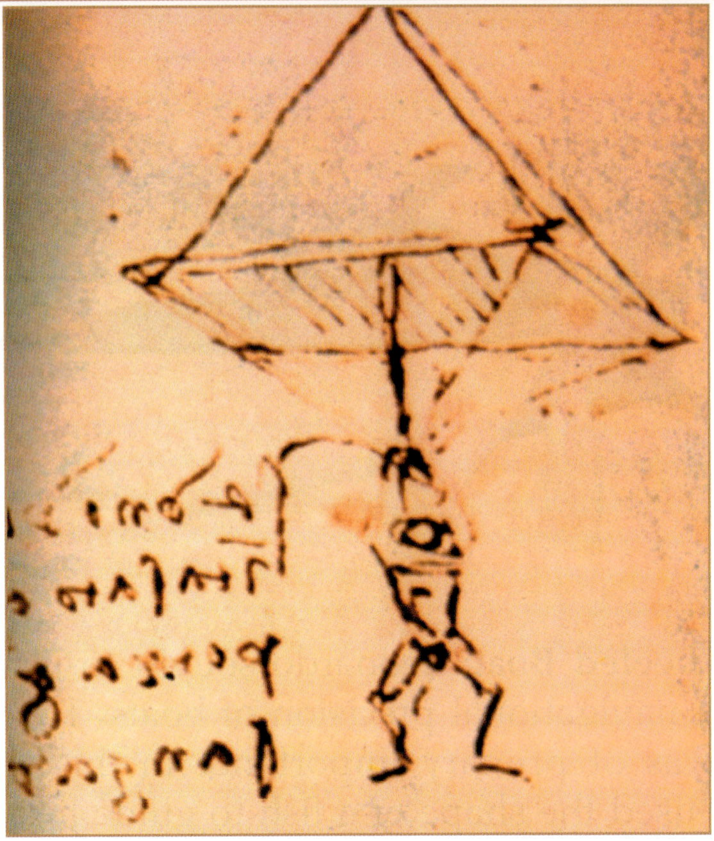

PARACHUTE
(Paracadute)
LV 22
Codex Atlantico f. 1058 v

PARACHUTE

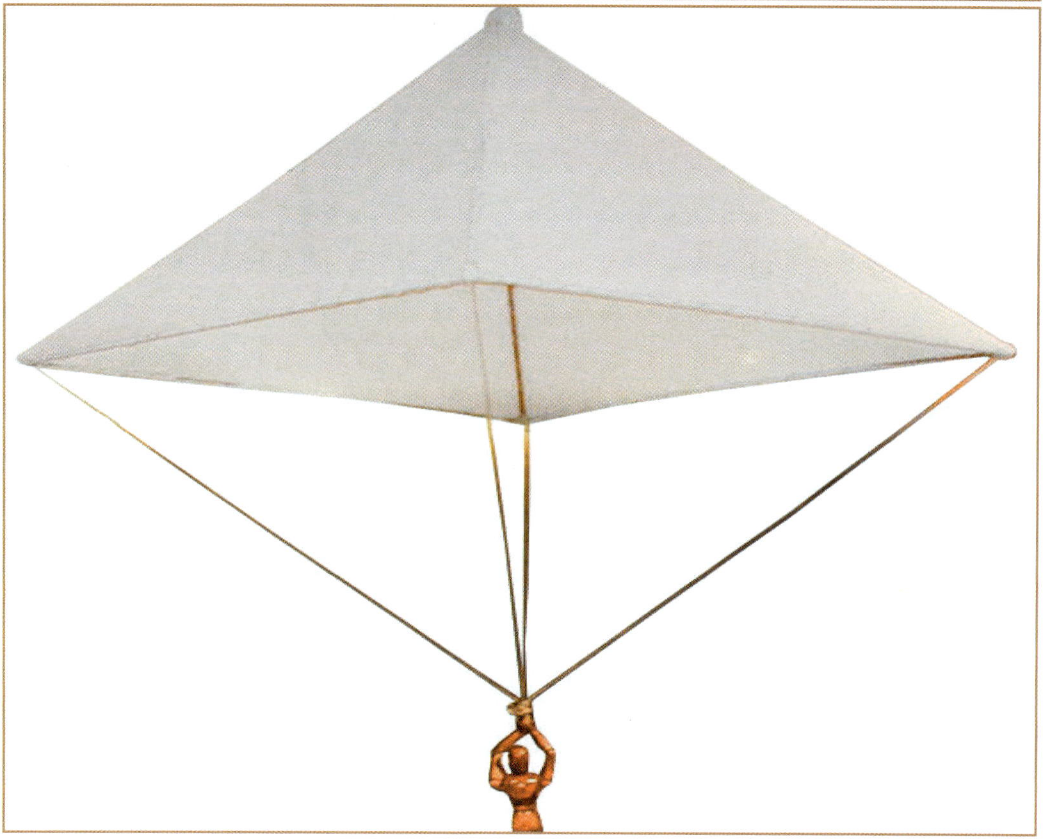

Parachute: Leonardo's concept for a parachute demonstrates how ingenious and futuristic his ideas were. He saw the parachute as a means of safely descending from great heights. He designed his to be made out of linen in the shape of a pyramid. It measured 7 meters per side, which is equivalent to the length of a limousine. Rather than include a hole at the top to promote stability — as is the case with modern parachutes — Leonardo relied on the porous nature of the material to achieve the same purpose.

STUDY OF A WING

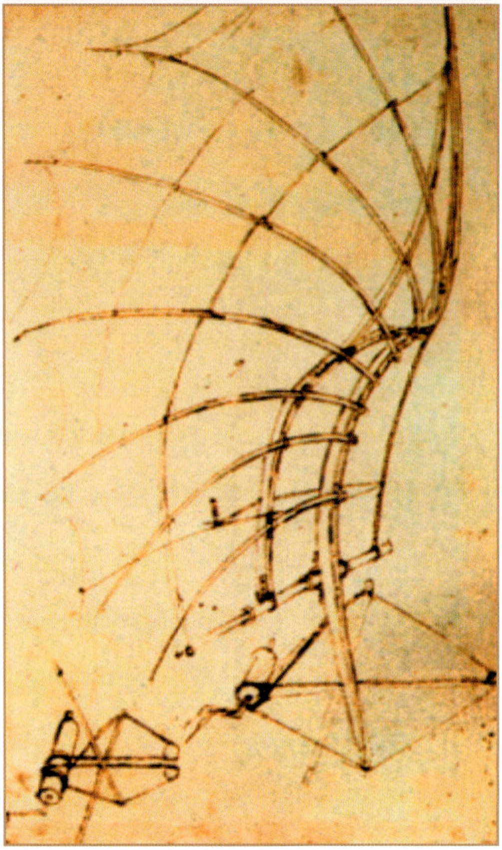

STUDY OF A WING
(Studio d'Ala Unita)
LV 53
Codex Atlantico f. 858 r

STUDY OF A WING

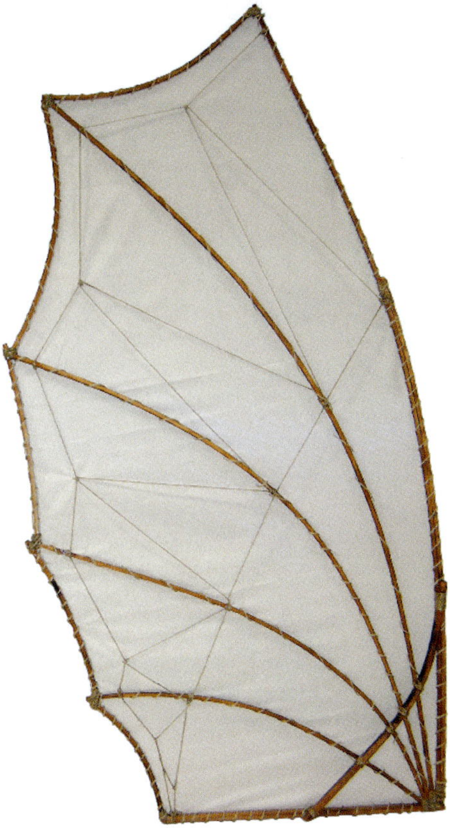

Study of a Wing: After analyzing the anatomy of birds, Leonardo decided to apply the results of his observations to his flying machines. In this study of a wing, he envisaged a single sheet of cloth stretched over a structure of wood and cane in the manner of the wing of a bat.

WATER SHOES

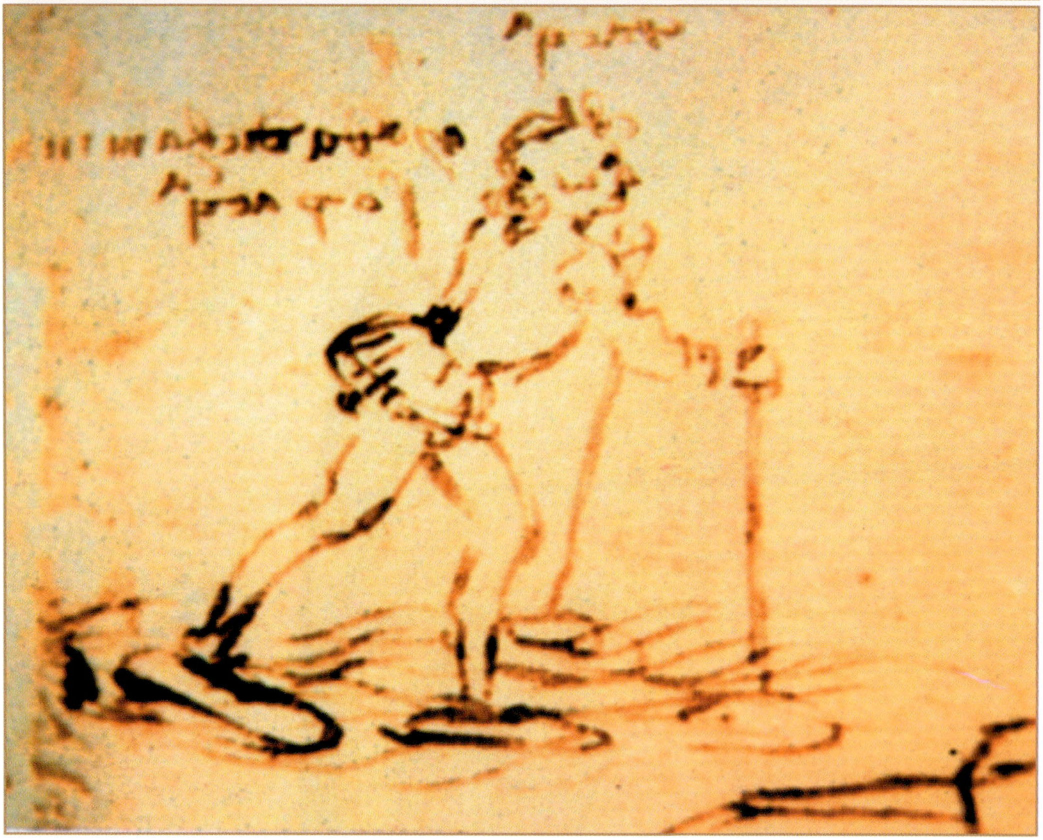

WATER SHOES
(Modo di Camminare sull'Acqua)
LV 28
Codex Atlantico f. 26 r

WATER SHOES

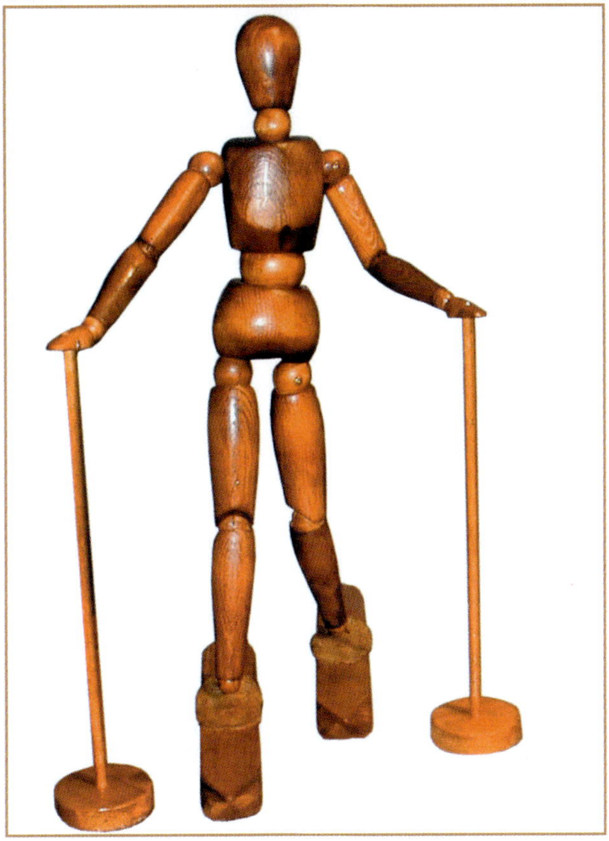

Water Shoes: Leonardo was fascinated by water and considered many mechanisms to help men dominate the element. It is well known that he was not the first who tried to create a mechanism to walk on water. This version illustrates Leonardo's concept for pontoons large enough to support a man's weight. The poles were intended to propel the person forward, much as a modern skier maneuvers on snow.

MORTAR ROUNDS

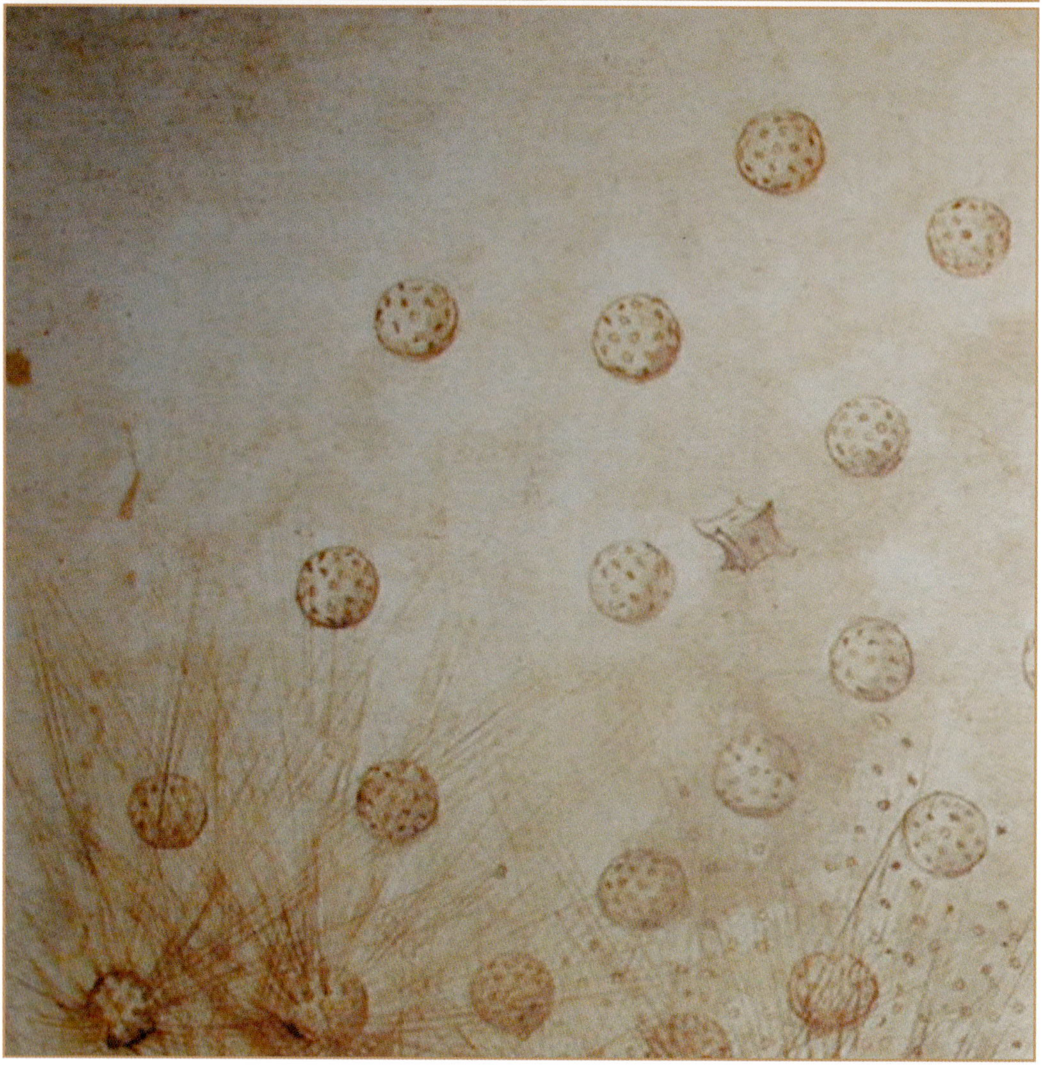

In this sketch, Leonard presents an idea for a new mortar shell that could do more destruction than conventional versions. Leonardo's shell was designed to break apart on contact with an object, generating hundreds of lethal fragments we now call shrapnel.

THE MACHINES OF THE MILITARY GALLERY

CATAPULT

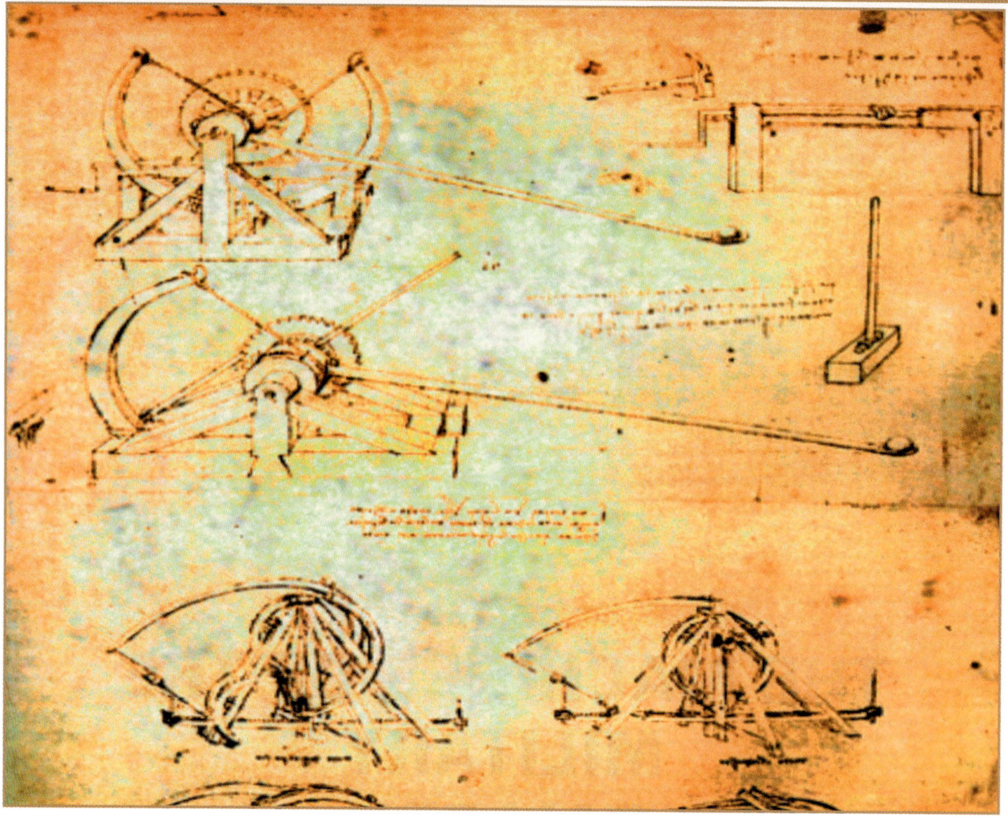

CATAPULT
(Catapulta)
LV 18
Codex Atlantico f. 140

CATAPULT

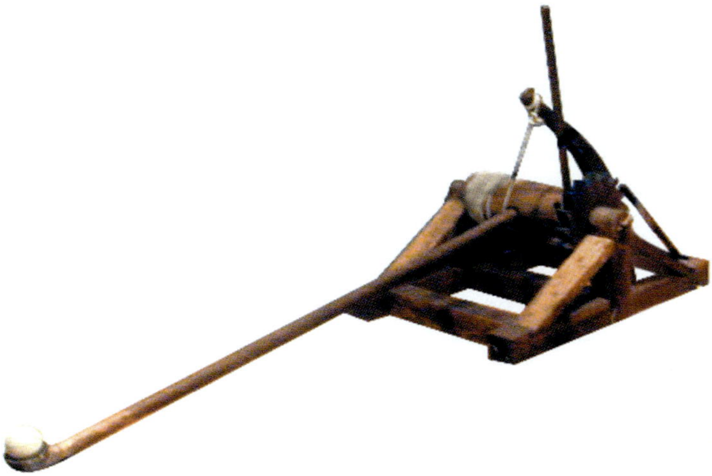

Catapult: This is one of the world's oldest known types of weapons. It was used to lob projectiles over the protective barriers of a fort or castle. Leonardo was interested in making them cast heavier shot over long distances more accurately. Many drawings with ideas to make a more powerful catapult can be found in the *Codex Atlantico*.

DEEP-SEA DIVER

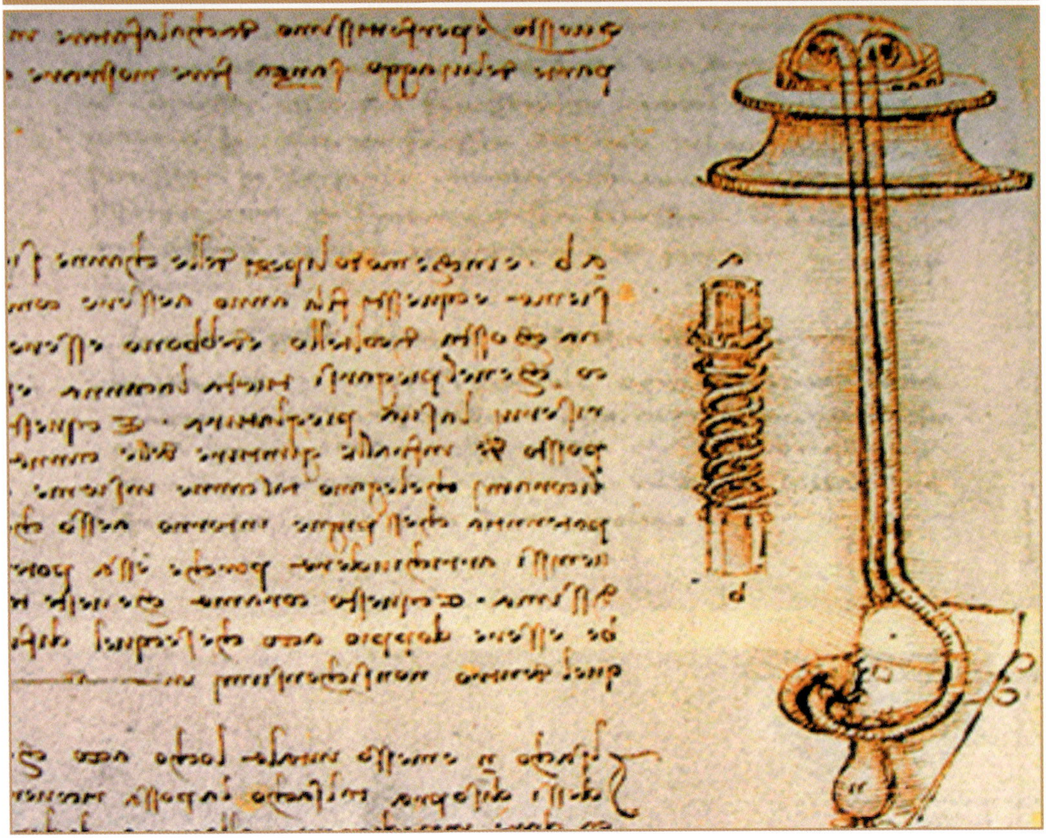

DEEP-SEA DIVER
(Palombaro)
LV 55
Codex Arundel f. 24 v

DEEP-SEA DIVER

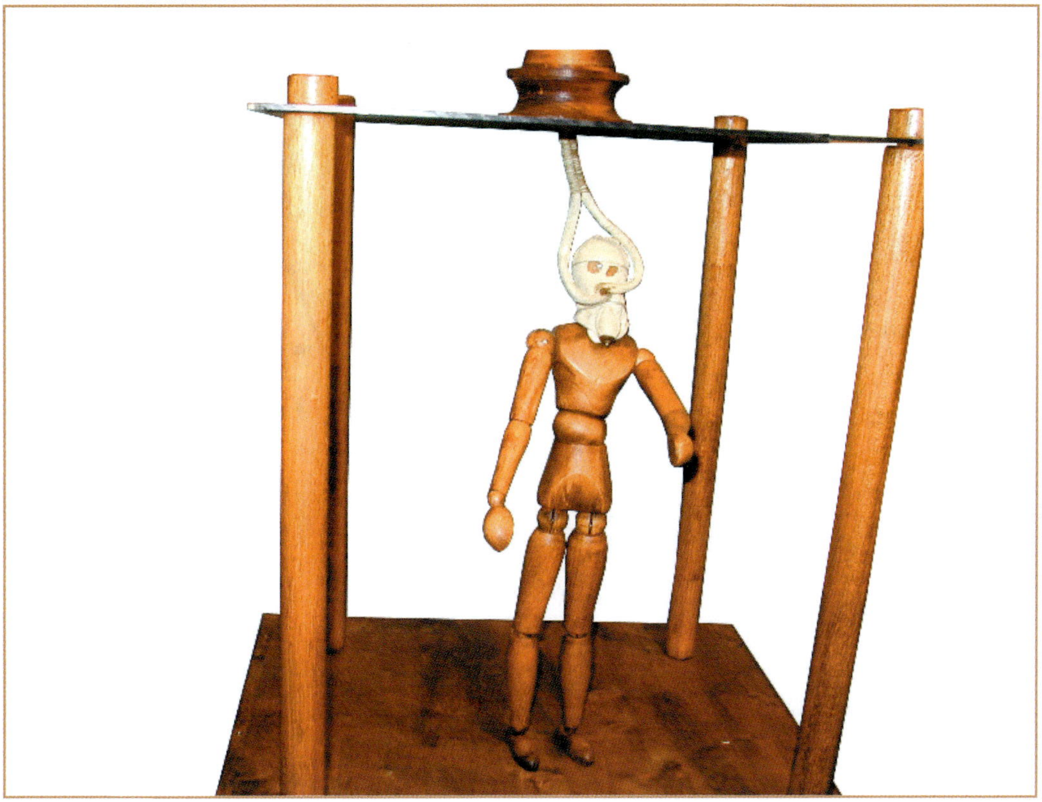

Deep-Sea Diver: Leonardo conceived of a way for men to move under water for considerable distances. Rather than expect divers to bring air from the surface directly down through a tube — something that could soon exhaust most — Leonardo reasoned that compressed air could be forced down the tube to reach the diver without extra effort on his part. Leonardo clearly anticipated the mechanisms used in modern underwater breathing equipment.

MOBILE STAIRS

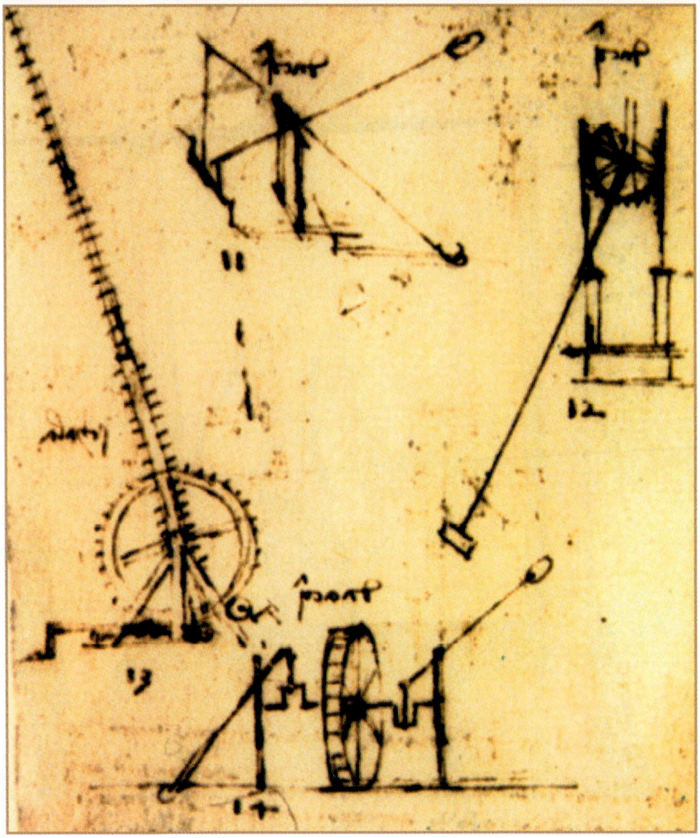

MOBILE STAIRS
(Scala Mobile)
LV 41
Codex Forster f. 46 v

MOBILE STAIRS

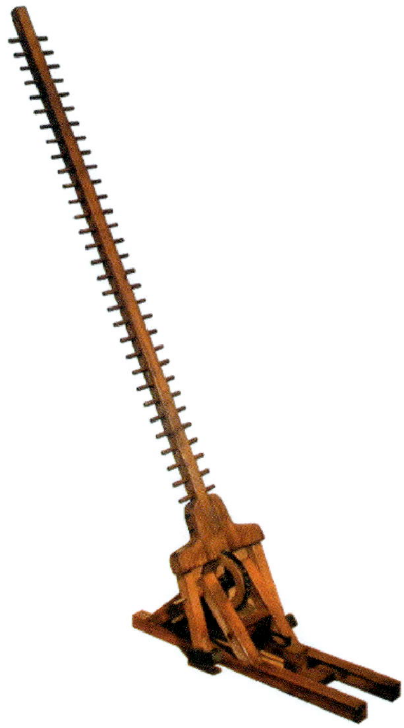

Mobile Stairs: This mobile staircase was designed to surmount a protective wall and constructed with gears and supports that made it more difficult for defenders to push away. From an engineering standpoint, the machine is interesting because it consists of removable parts connected to a toothed, worm screw mechanism, permitting it to be raised and lowered as needed from one place to another.

MORTAR BOAT

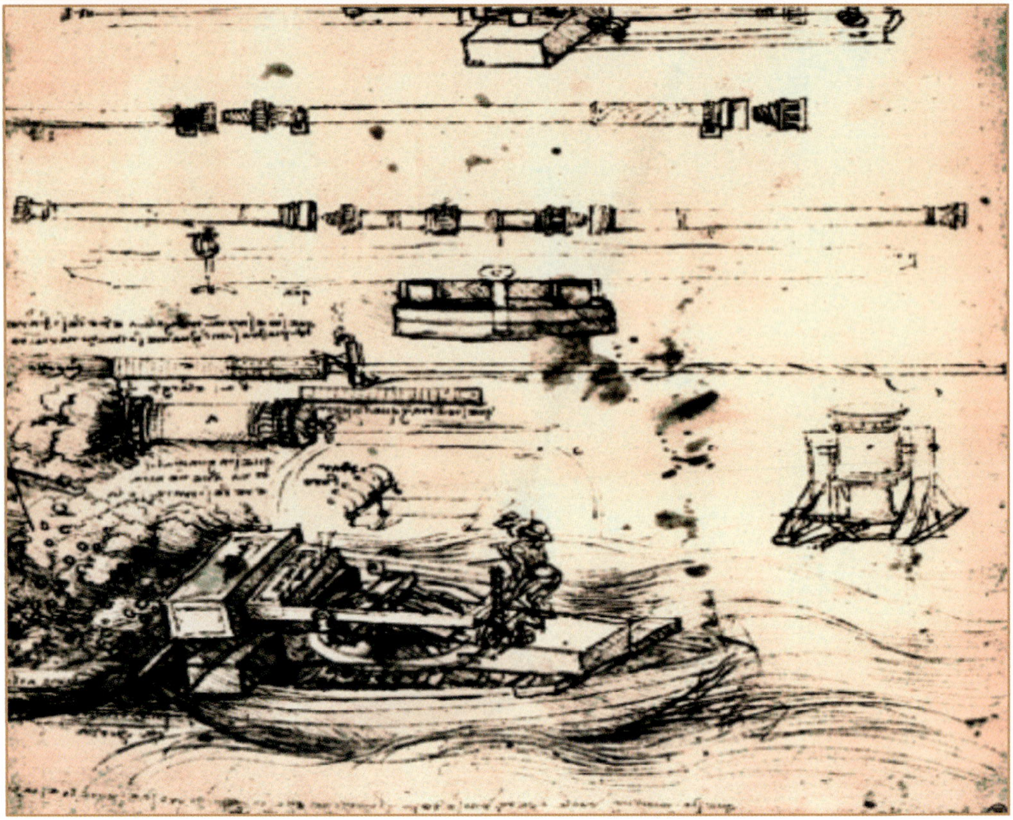

MORTAR BOAT
(Cannone Navale)
LV 15
Windsor B.R. n. 12652

MORTAR BOAT

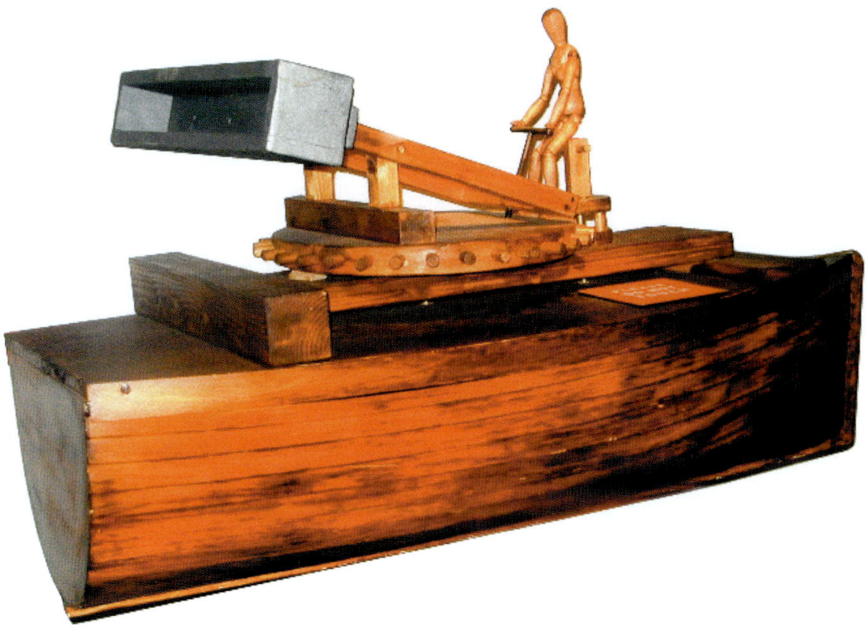

Mortar Boat: Naval warfare became one of Leonardo's principal areas of interest. He proposed equipping special boats with revolving platforms to fire cannons or mortars in all directions. In this example, he envisaged a large square mortar to fire incendiary shells to create fires on enemy ships.

MOVABLE BRIDGE

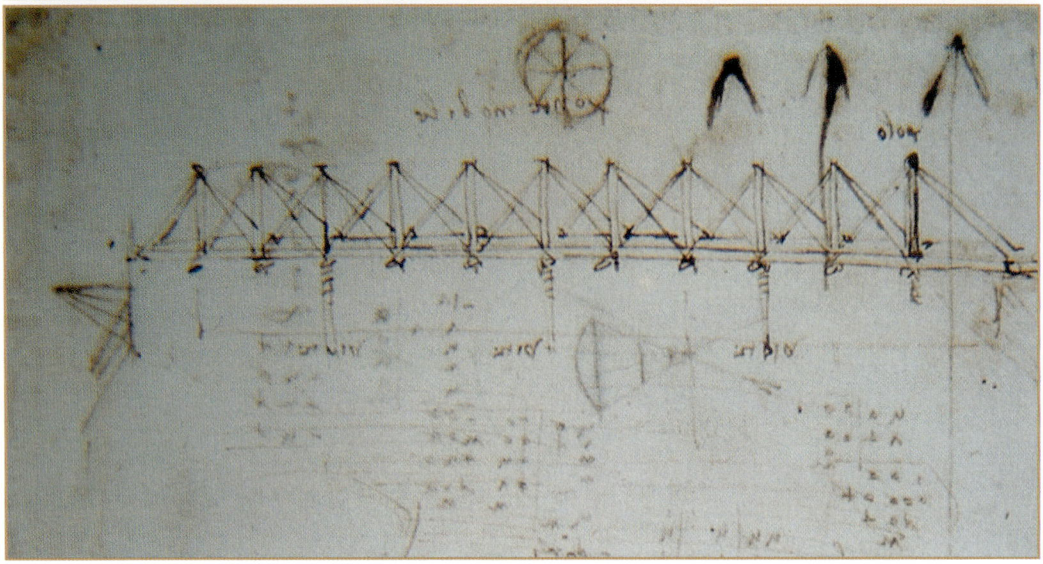

MOVABLE BRIDGE
(Ponte Mobile)
LV 26
Codex Atlantico f. 855 r

MOVABLE BRIDGE

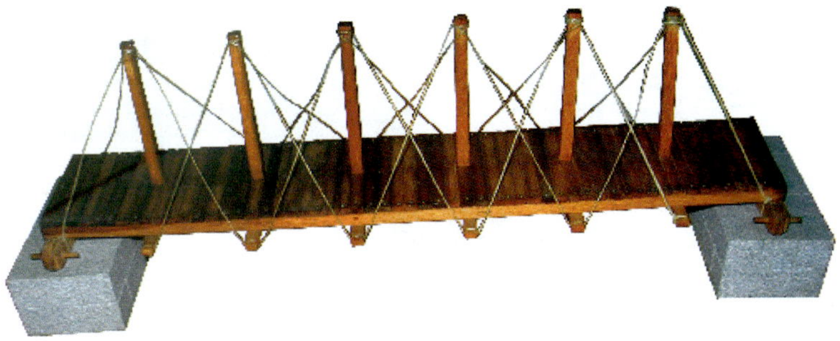

Movable Bridge: In a 1482 letter to Duke Ludovico Sforza of Milan, Leonardo boasted that he could build bridges that would be extremely lightweight, but strong — characteristics potentially useful during wartime. The bridges were designed to be rapidly assembled and disassembled. The bridge shown here is mobile at both ends, permitting soldiers to cross the defensive moat of a fortress but move it away quickly so the other side could not use it themselves in a counterattack.

PORTABLE BRIDGE

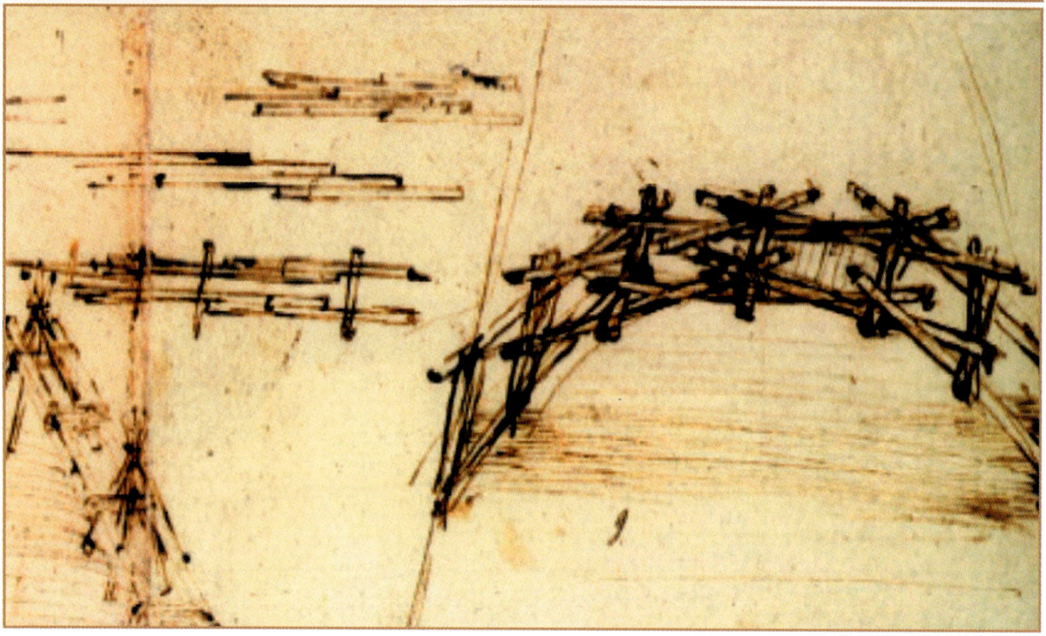

PORTABLE BRIDGE
(Ponte Arcuato)
LV 38
Codex Atlantico f. 22 r

PORTABLE BRIDGE

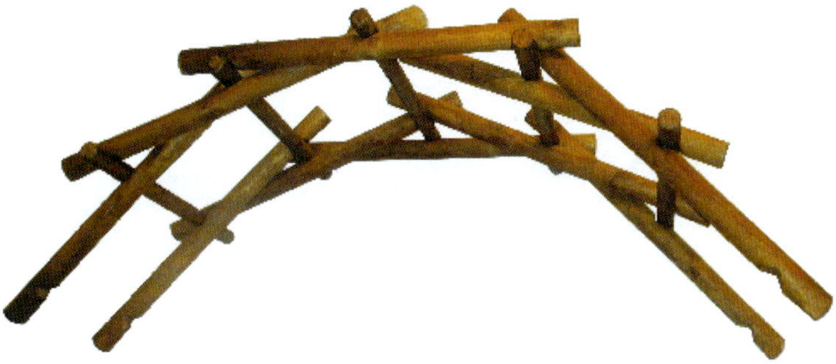

Portable Bridge: This is another of the "light and strong bridges" that Leonardo promised to design and build in his letter to the Duke of Milan. In Leonardo's conception, they could be built rapidly with materials that would be easy to find and carry, such as small logs and ropes. This bridge, however, is built entirely of interlocking logs, without any ropes or hardware. It is strong enough to support the weight of several men crossing it simultaneously. It permitted rapid and unexpected troop movements across rivers and ravines, thus contributing to the element of surprise.

SCORPION

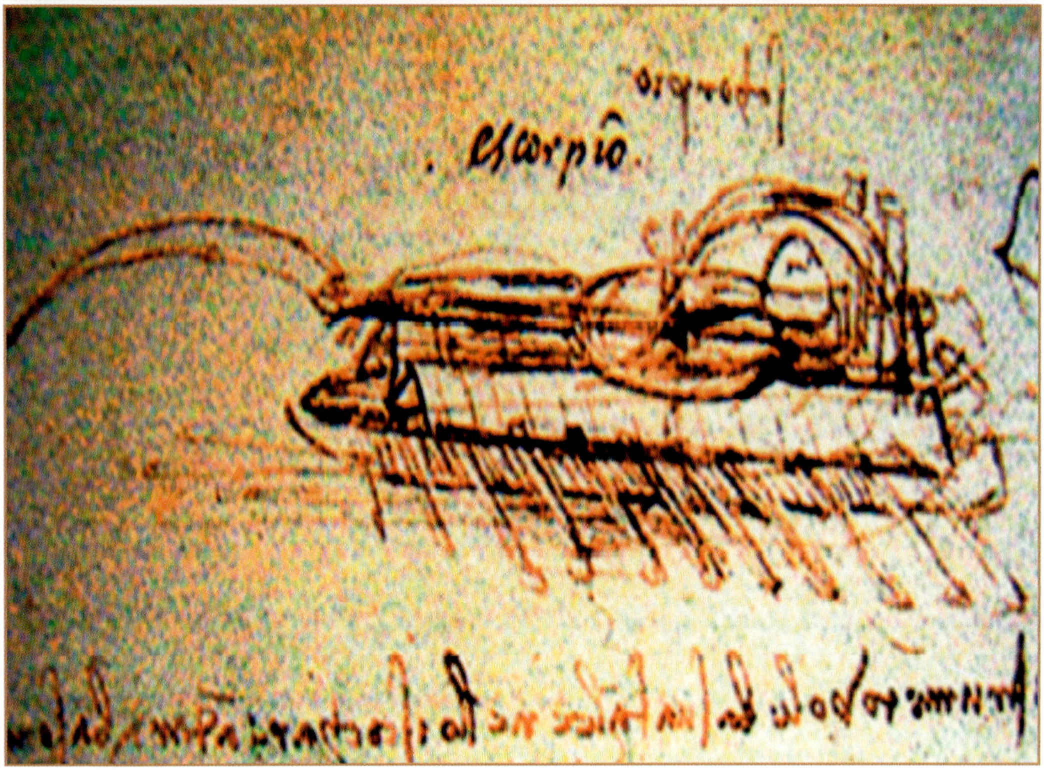

SCORPION

(Barca con Falce – Escorpio)

LV 43

Codex Ashburnham 2037 f. 8 r

SCORPION

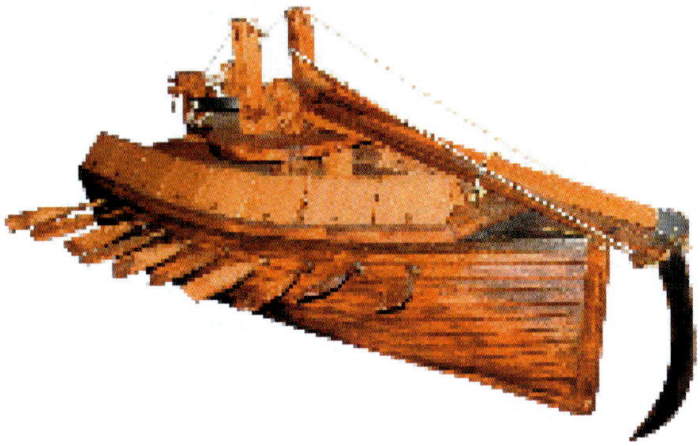

Scorpion: The Scorpion was designed to slash the sails of opposing ships. Its revolving platform would permit it to maneuver and drop the large scythe where it could do the most damage. Note how Leonardo shielded the positions of the Scorpion's rowers with screens to protect them from the battle above.

SCORPION — A MODERN USE

Scorpion — A Modern Use: One of the modern uses for a key feature of Leonardo's Scorpion warship is found in the way a log crane functions. The capability of the crane's arms to twist, flex, and be moved to lift a log in whatever position the operator might find it, mirrors Leonardo's concept that the scythe had to be maneuverable and flexible to inflict maximum damage on the sails of opposing ships.

SCORPION — A MODERN USE

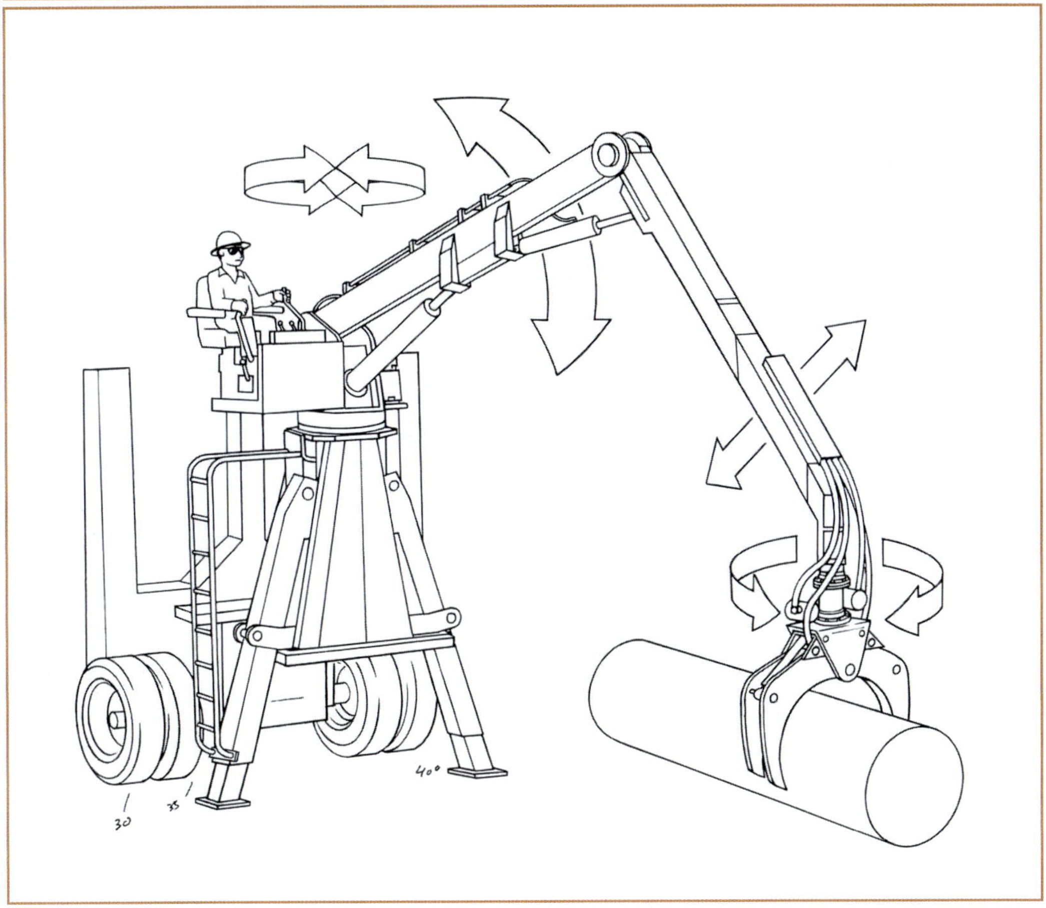

Illustration
Jamie Pfeifer Graphics

SHELLS

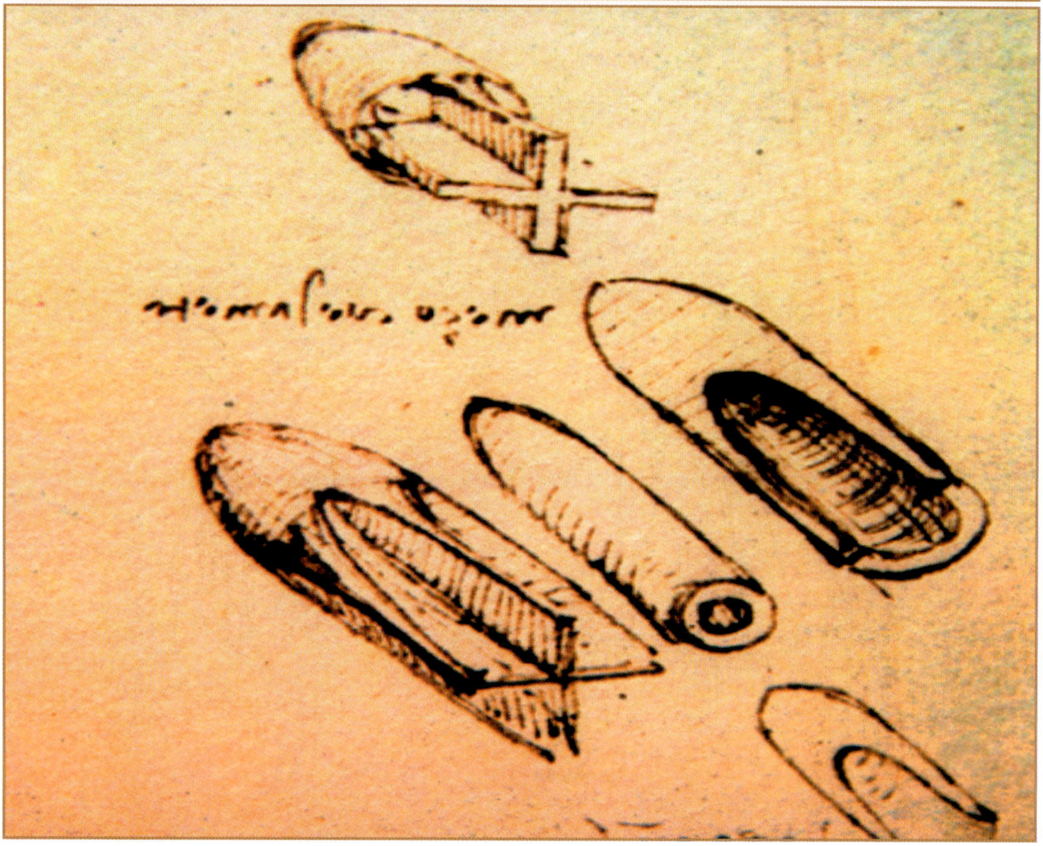

SHELLS
(Proiettili Ogivali)
LV 16
Codex Arundel f. 54 r

SHELLS

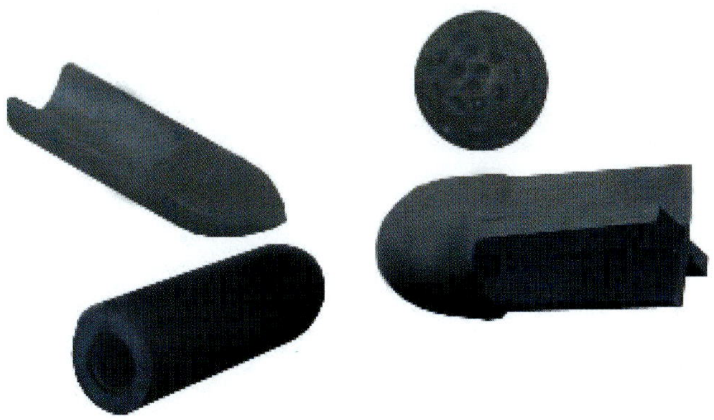

Shells: Leonardo came to understand how air resistance affected the trajectory of cannon balls and mortar rounds. He experimented with jets of water to simulate air currents. He addressed the problem by designing a variety of projectiles with shapes and tails that are amazingly modern in appearance.

TANK

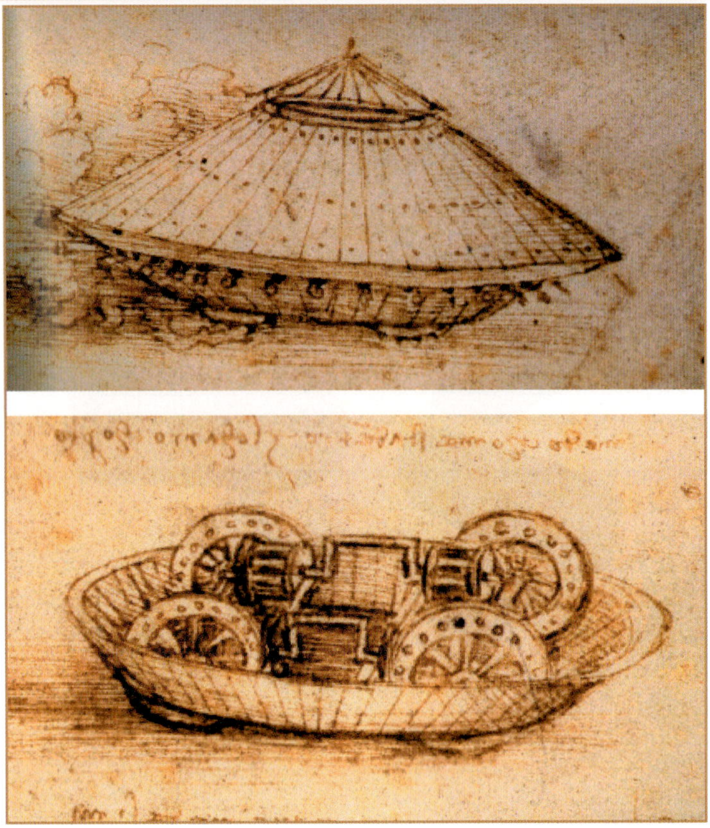

TANK
(Carro Armato)
LV 57
British Museum f. 1030

TANK

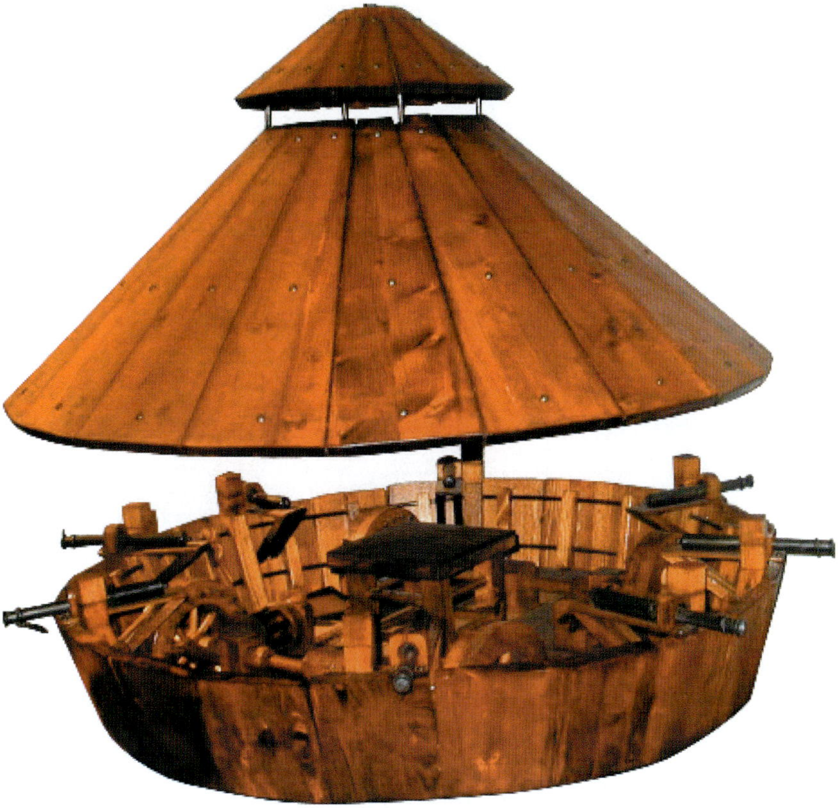

Tank: The idea of a covered military vehicle that could penetrate enemy lines in advance of soldiers on foot first appeared in the Middle Ages. Leonardo conceived of such a vehicle in the shape of a tortoise with cannons positioned all around the inside. Some people had thought of moving such vehicles with sails, but Leonardo's solution was a system of wheels worked by eight men. He even thought of using horses *inside* the vehicle, but the risk of the animals' panicking in such a tight, noisy space made him change his mind.

WALL-DEFENSE SYSTEM

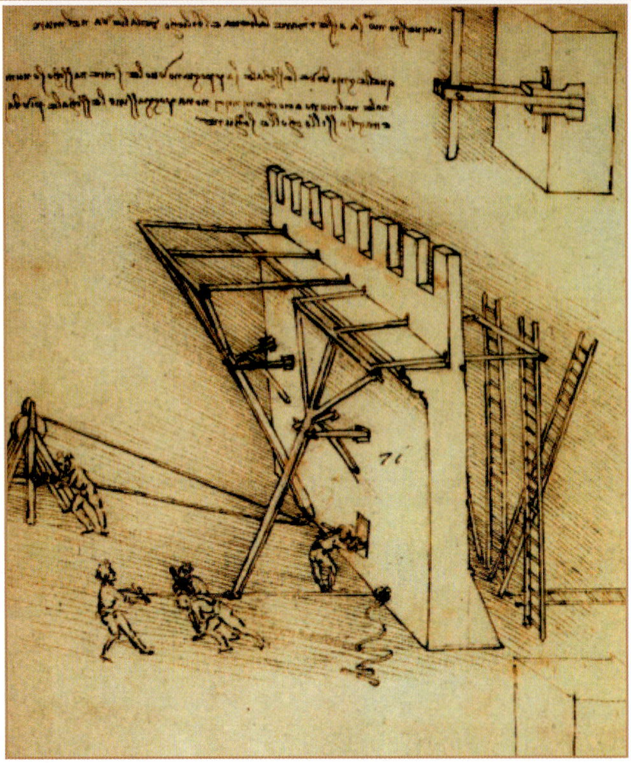

WALL-DEFENSE SYSTEM
(Difesa Delle Mura)
LV 44
Codex Atlantico f. 139 r

WALL-DEFENSE SYSTEM

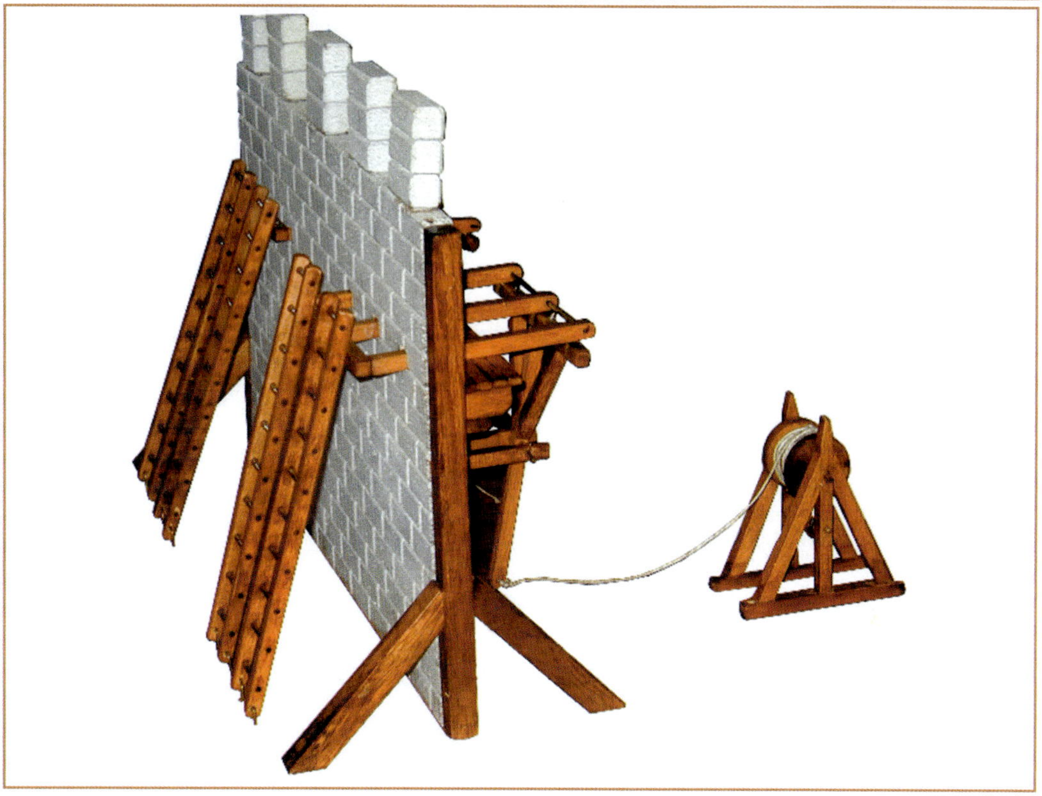

Wall-Defense System: This is an ingenious system for dislodging enemy ladders. If soldiers managed to prop their ladders against the walls of a fort or palace, long rods attached to a crank-operated lever would push the ladders backwards as diagrammed below.

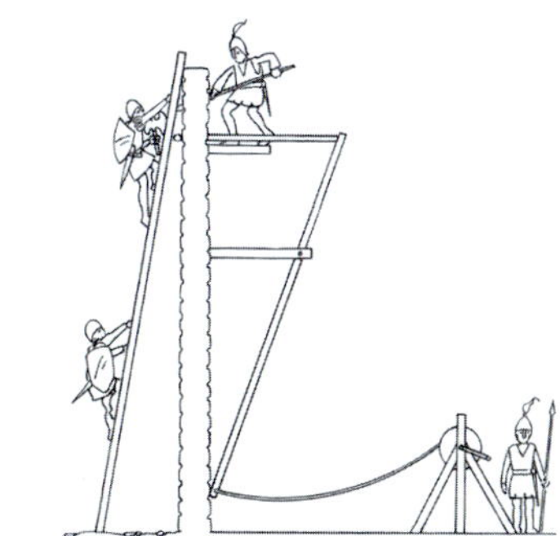

Illustration:
Jamie Pfeifer Graphics

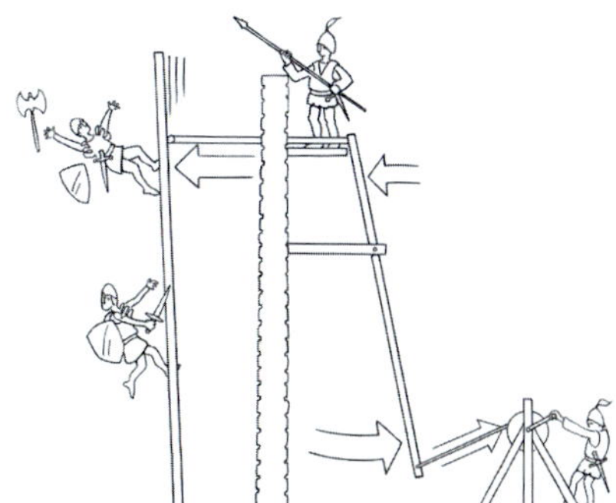

WALL-SCALING TECHNIQUE

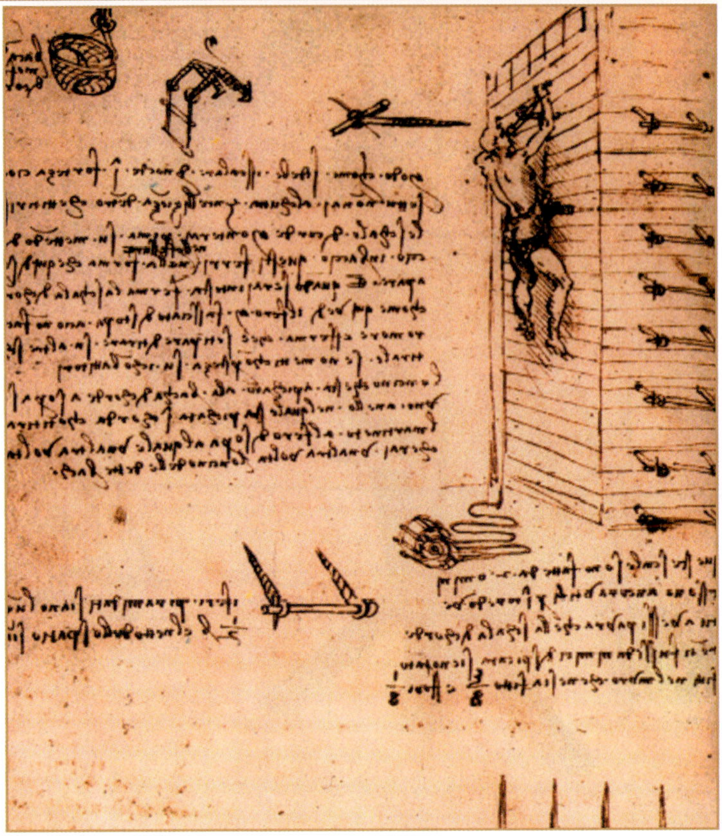

WALL-SCALING TECHNIQUE
LV 45
Codex Atlantico f. 59 v

WALL-SCALING TECHNIQUE

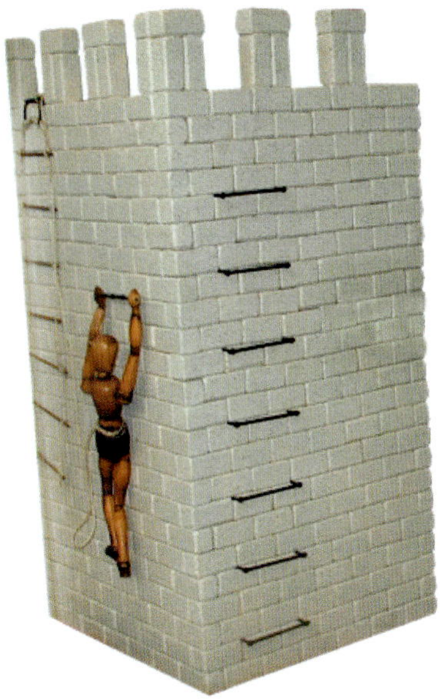

Wall-Scaling Technique: Leonardo invented many attack and defense systems. Here a man is depicted assaulting a vertical fortress wall by placing and recovering spikes driven into the surface of the wall. Variations on this technique are now widely employed in mountain and rock-climbing activities.

WAR WAGON

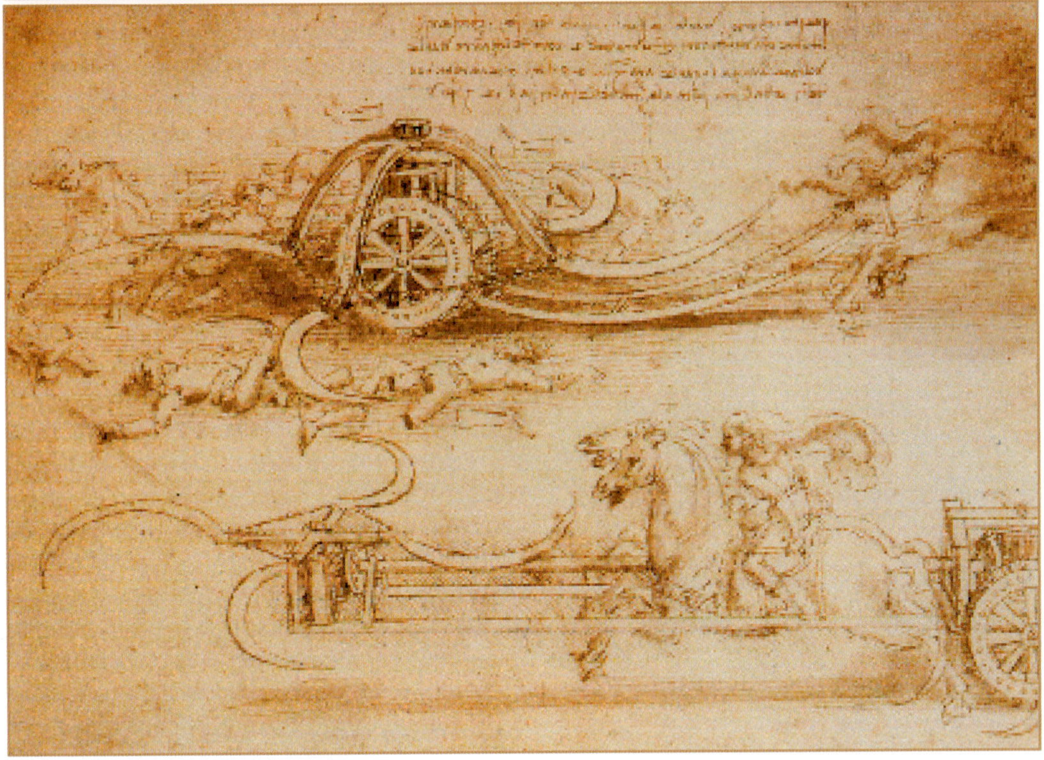

WAR WAGON
(Carro falciante)
LV 42
Bibliteca reale Torino f. 14 r (15583)

WAR WAGON

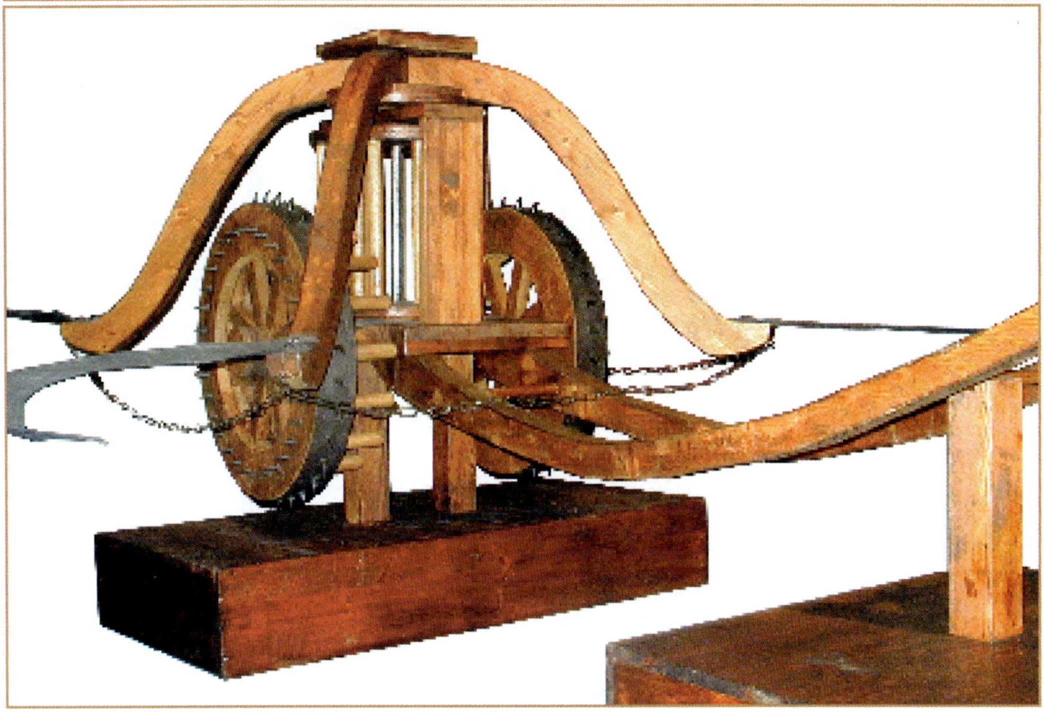

War Wagon: This type of reaping wagon was known to have existed in Roman days. As Leonardo mentioned in his notes, it could be quite destructive, requiring the opposition to use: " ... archers, men with slings, stones and guns ... " to defend against it. Leonardo's design has a wagon pulled by horses with rotating scythes set in motion by an ingenious gearing mechanism. This one, like other machines, would not work as Leonardo drew and described it. He always kept a key element of an important machine to himself for security reasons.

THE DA VINCI EXPERIENCE

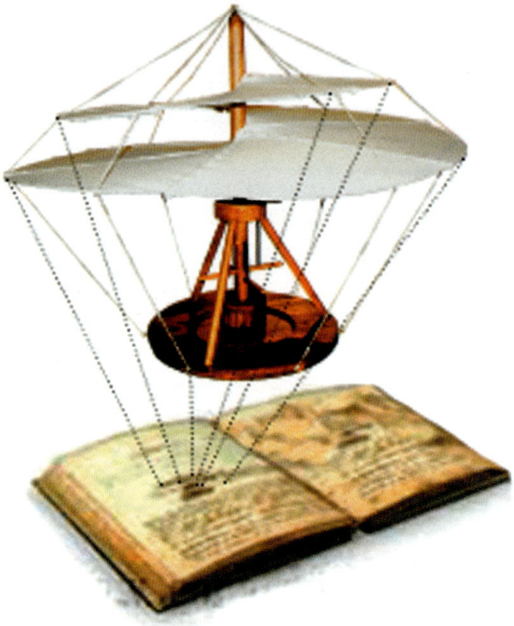

Machines Designed and Described by
LEONARDO da VINCI

THE MACHINES OF THE MECHANICAL GALLERY

ARCHIMEDES' SCREW

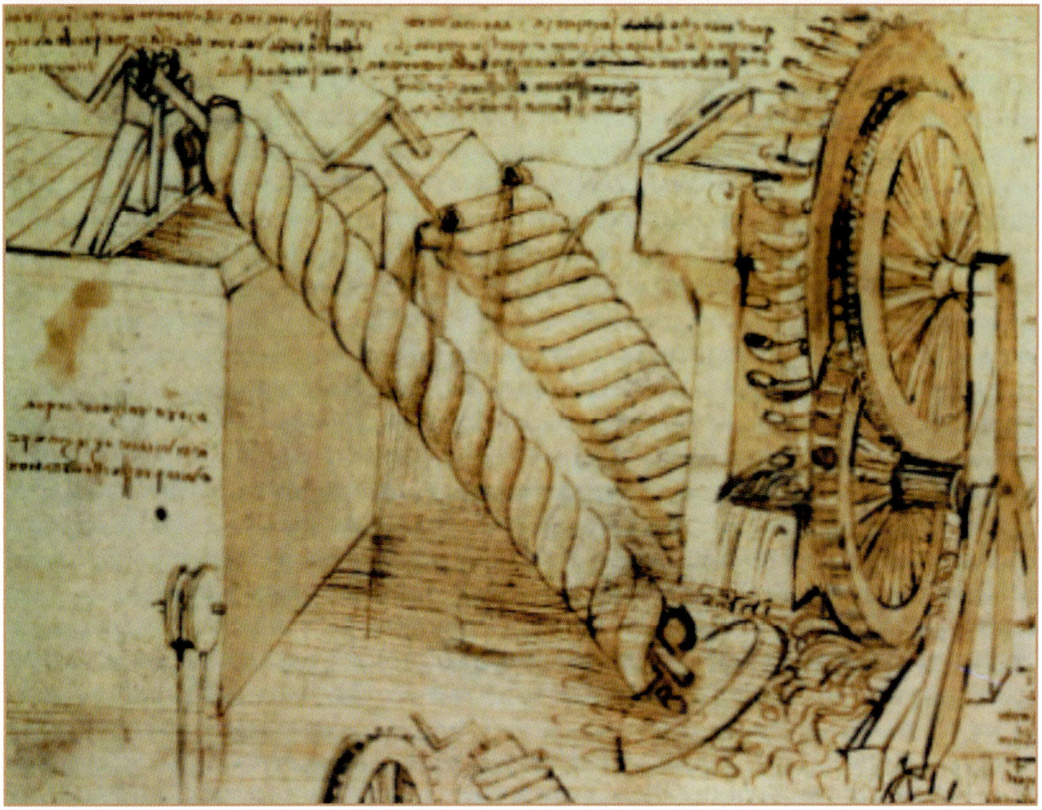

ARCHIMEDES' SCREW

(Vite d'Archimede)

LV 02

Codex Atlantico F. 26 v

ARCHIMEDES' SCREW

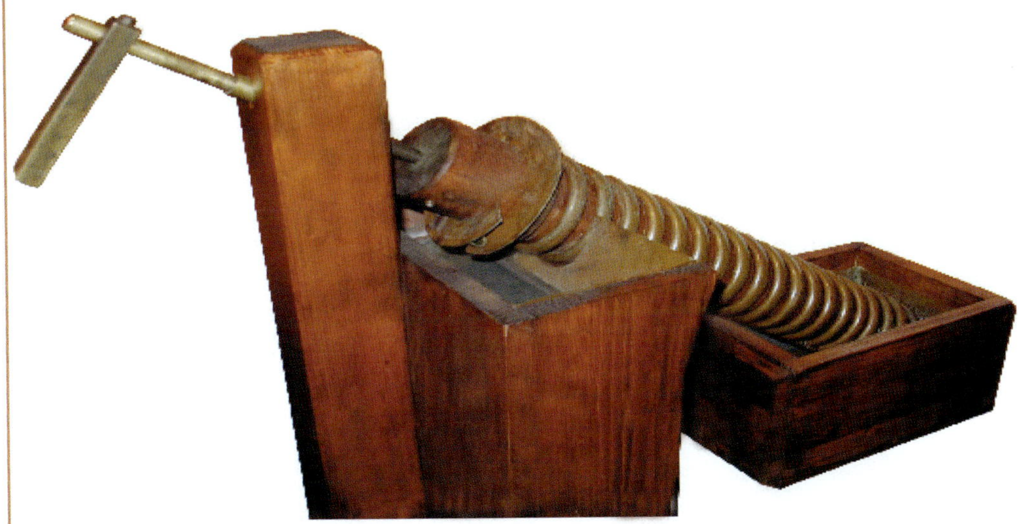

Archimedes' Screw: This machine, designed to lift water from a river to a field above, was not one of Leonardo's inventions. It was known in ancient Egypt and was later studied in Greek times by Archimedes. During the Renaissance every engineer was familiar with it and put it to use on a regular basis. Leonardo drew a great number of plans for these water screws throughout his career, often adding ingenious improvements to make them function more efficiently.

AUTOLOCK MECHANISM

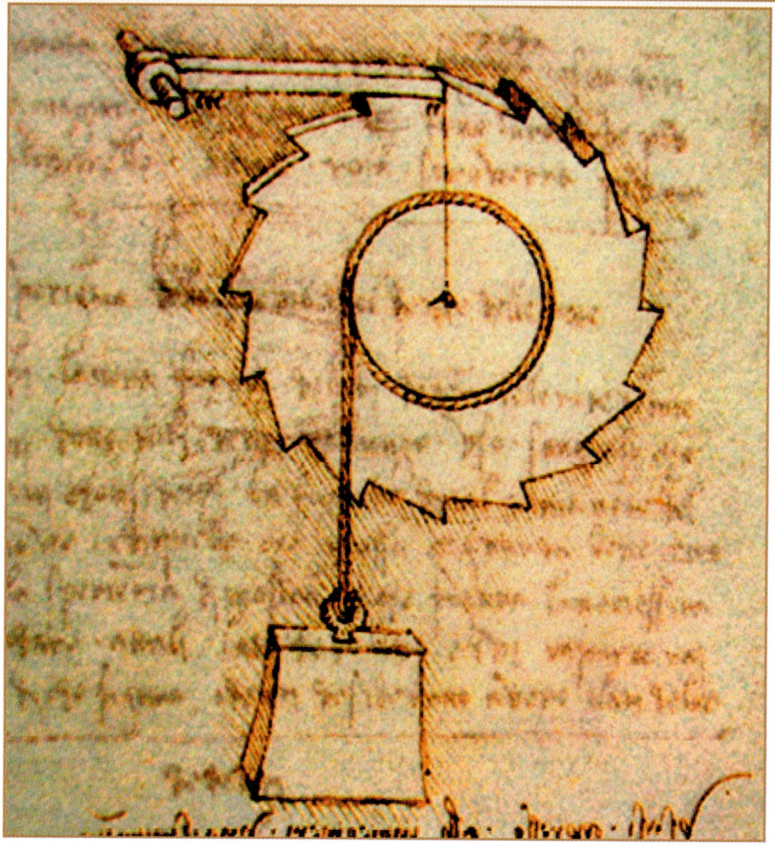

AUTOLOCK MECHANISM
(Meccanismo Autobloccante)
LV 68
Codex Madrid I f.97 r

AUTOLOCK MECHANISM

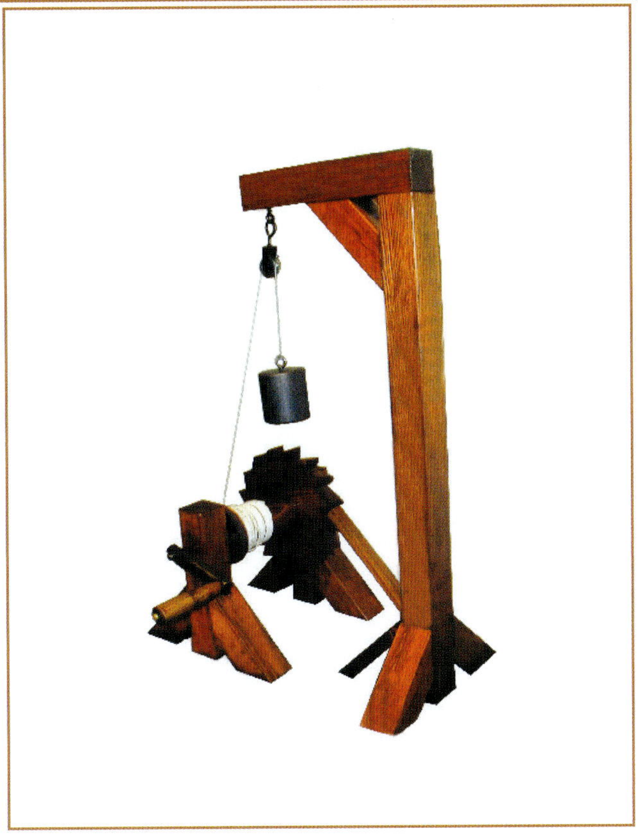

Autolock Mechanism: The use of this device was essential in machines designed for lifting great weights. It consisted of a wheel with teeth set at an acute angle and a flat-tipped hook that engaged the teeth one at a time so that the wheel could not turn backwards. The same principle makes elevators safe in tall buildings.

BALL BEARINGS—A

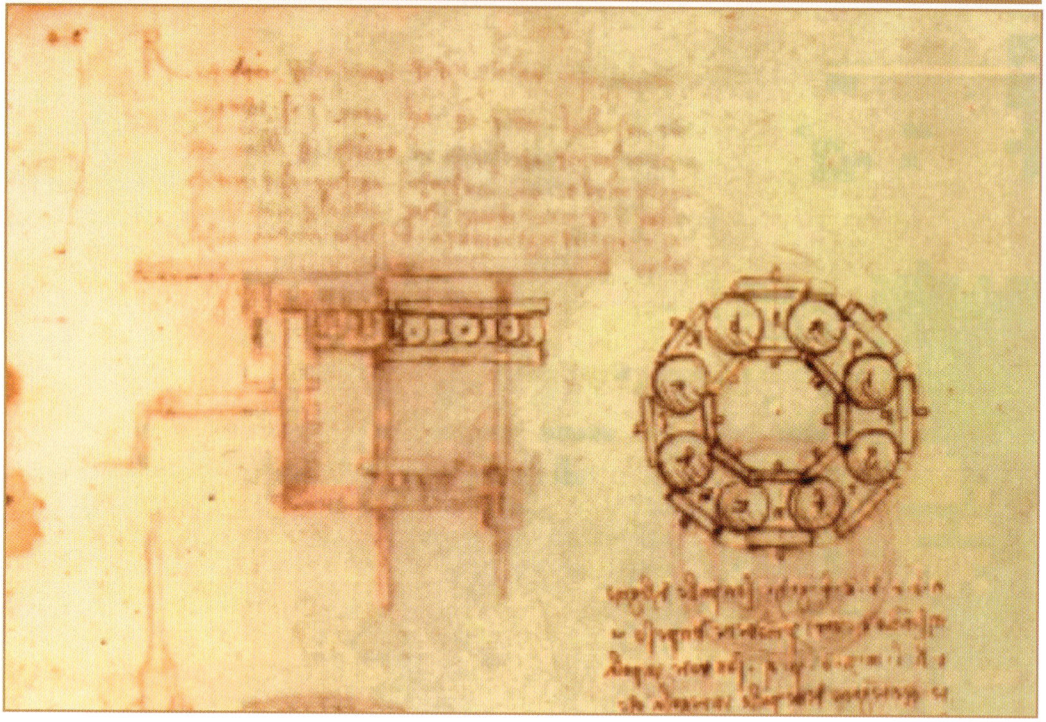

BALL BEARINGS
(Cuscinetto a Sfere)
LV 10
Codex Madrid I f. 20 v

BALL BEARINGS—A

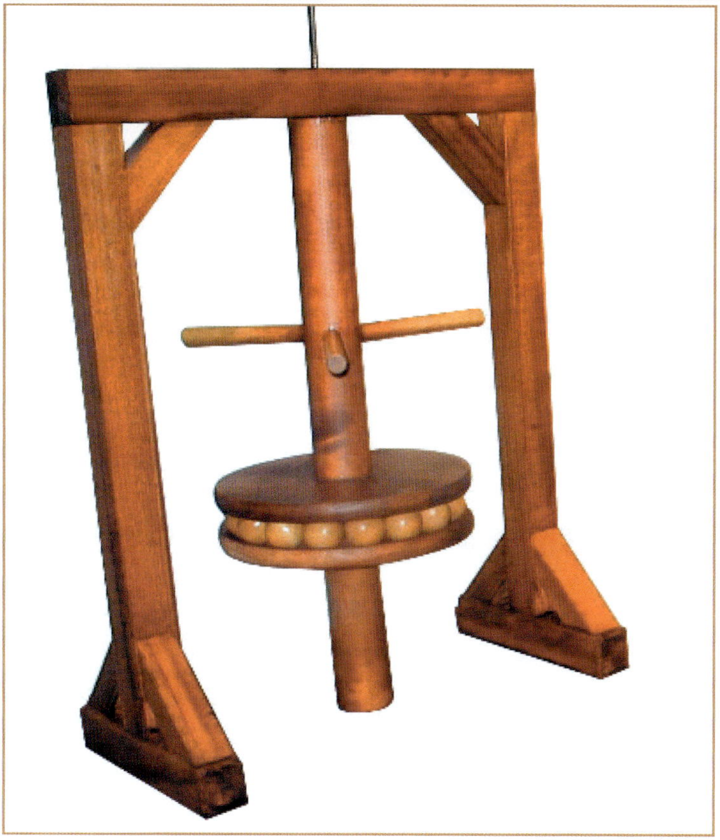

Ball Bearings: Leonardo recognized that when two wooden plates rub against each other during a machine's operation, friction and heat would soon wear one or both plates out. The use of freely-rotating balls between two plates dissipates the friction and heat among many continually changing surfaces, thus reducing the wear and tear that otherwise would occur.

BALL BEARINGS—B

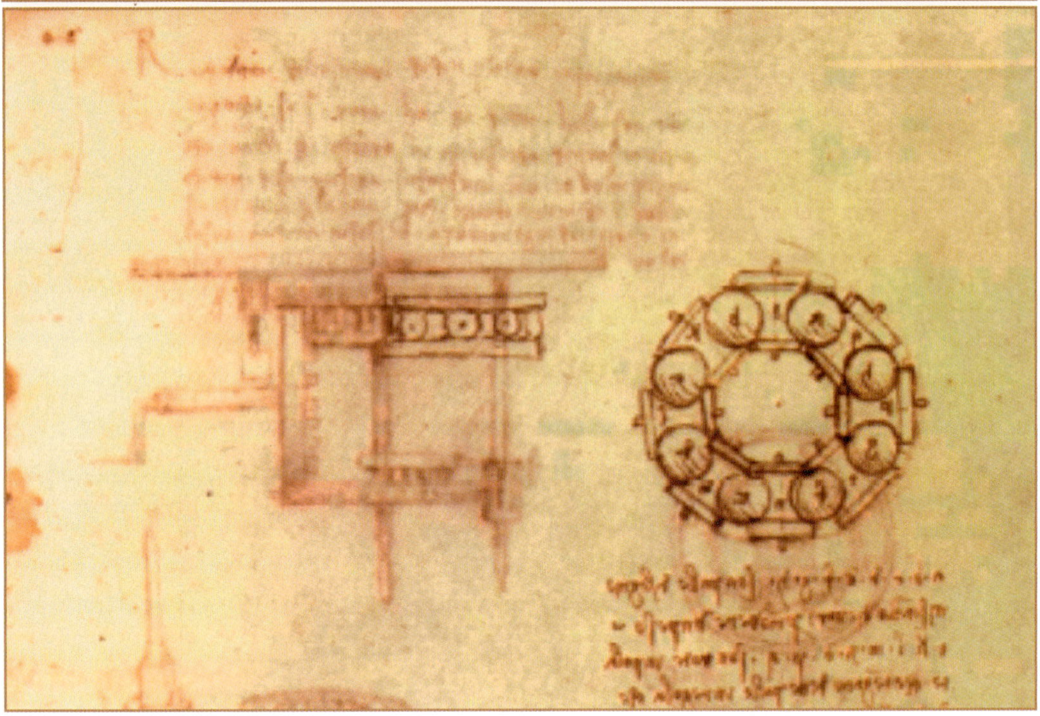

BALL BEARINGS
(Cuscinetto a Sfere)
LV 10
Codex Madrid I f. 20 v

BALL BEARINGS—B

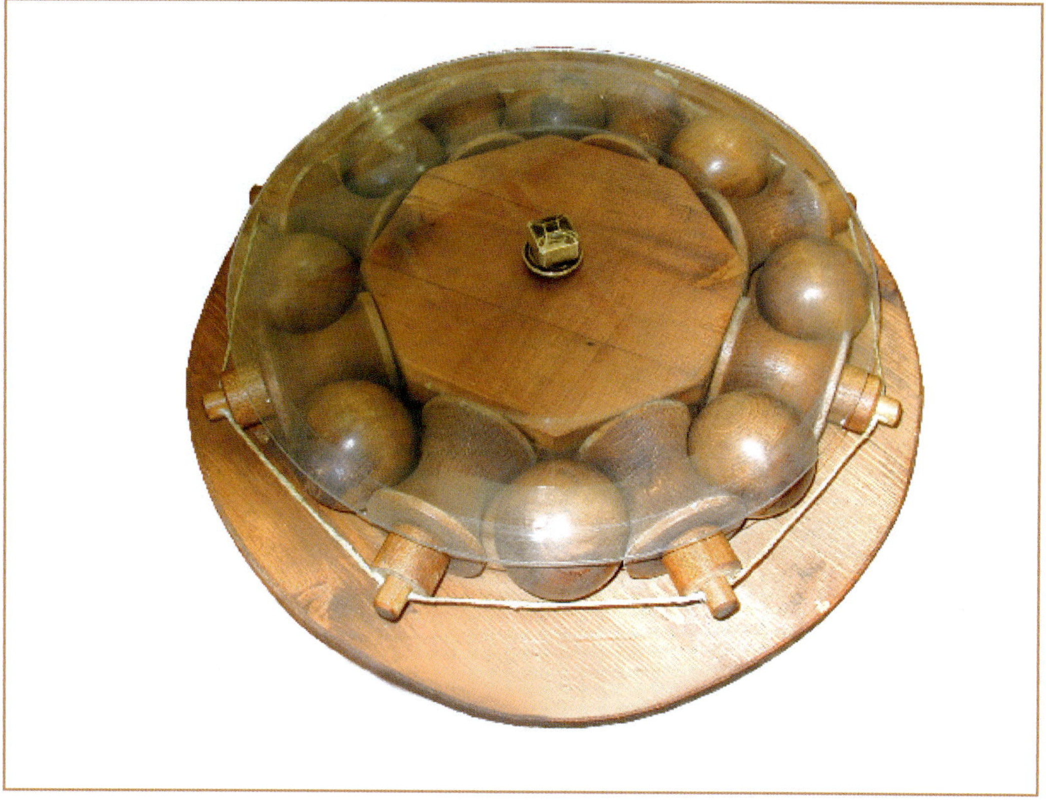

Ball Bearings: Leonardo sought different ways to overcome the effects of friction on his machines. Here, concave-sided spindles, rotating on their own axes, prevent lateral movement by the ball bearings while still allowing them to rotate freely. The ball-and-spindle system reduces friction, allowing the top platform to turn easily even when carrying a heavy load.

CAM HAMMER

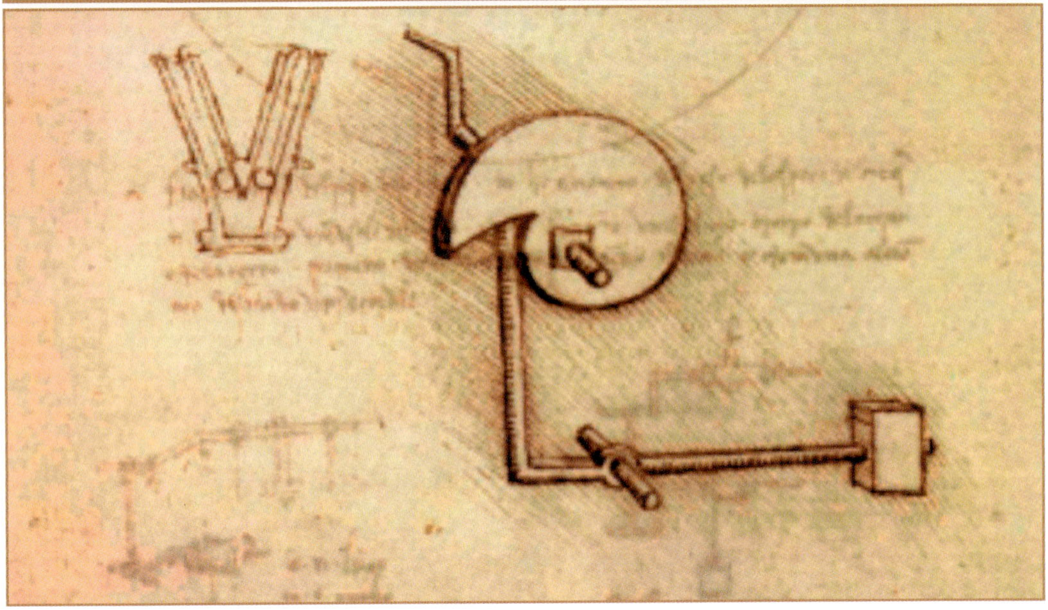

CAM HAMMER
(Martello a Camme)
LV 30
Codex Madrid I f. 6 v

CAM HAMMER

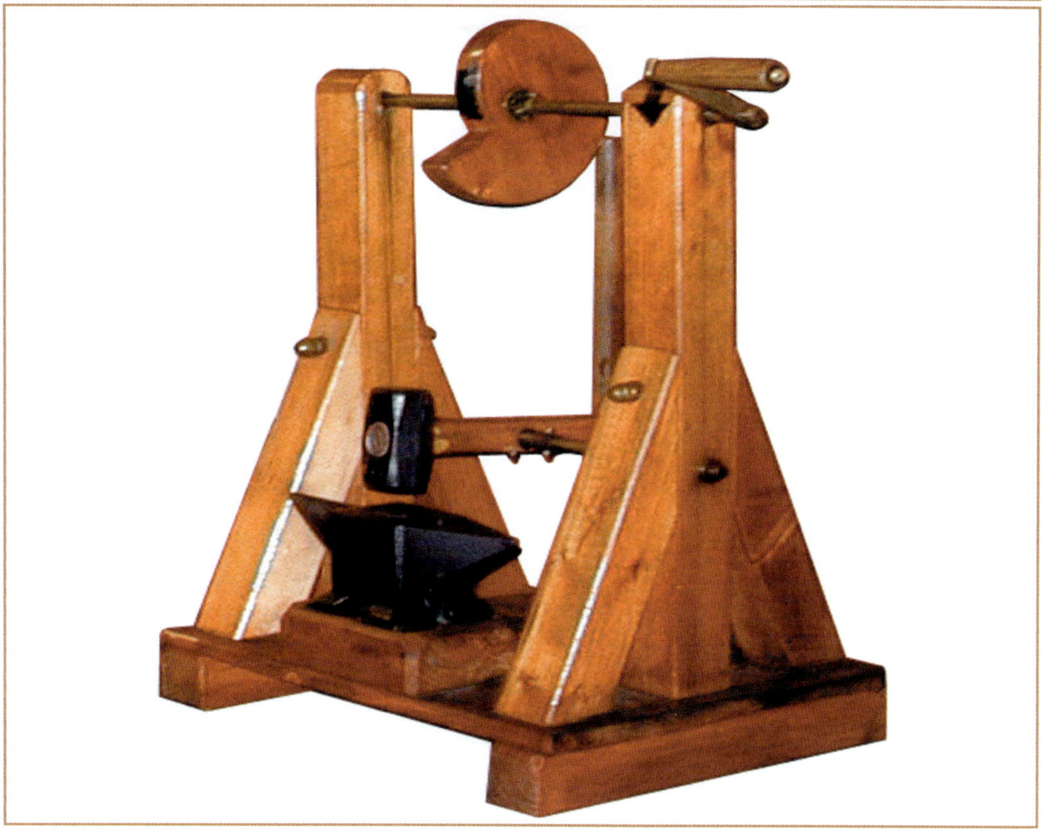

Cam Hammer: Leonardo often made use of cams — the projection on a rotating rod that generates either an even reciprocating motion or a regular percussive action. In this case, the cam raises the head of the hammer smoothly before releasing it to drop on a target with force. This concept is prevalent in the operation of many modern machine tools.

CHAMBER OF MIRRORS

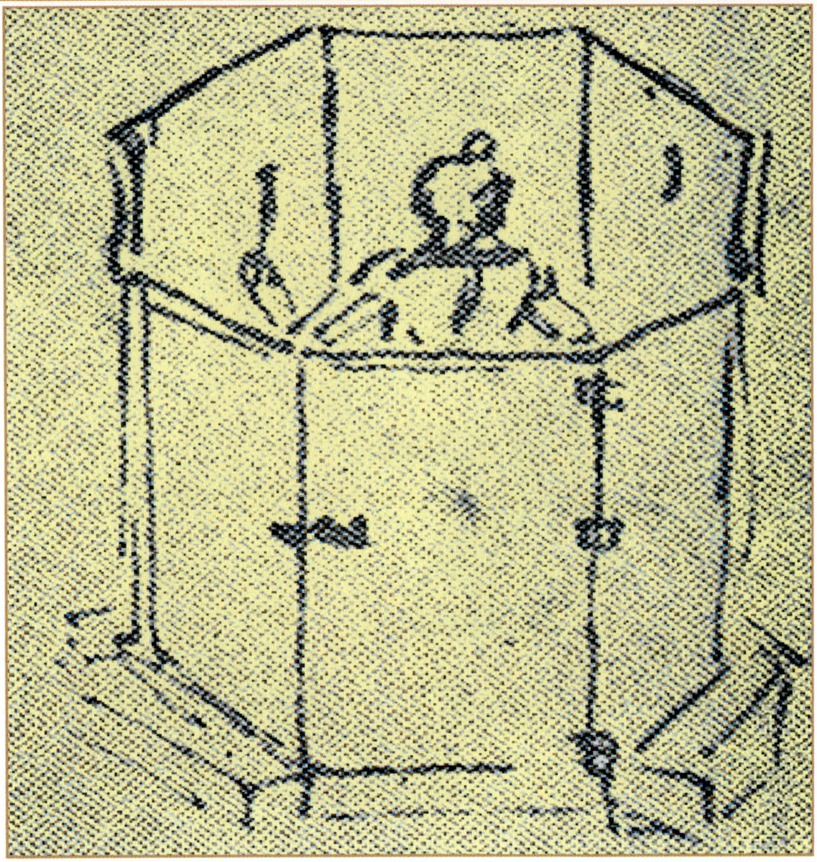

CHAMBER OF MIRRORS
LV32
(Camera degli specchi)
Manoscritto B f. 28 r

CHAMBER OF MIRRORS

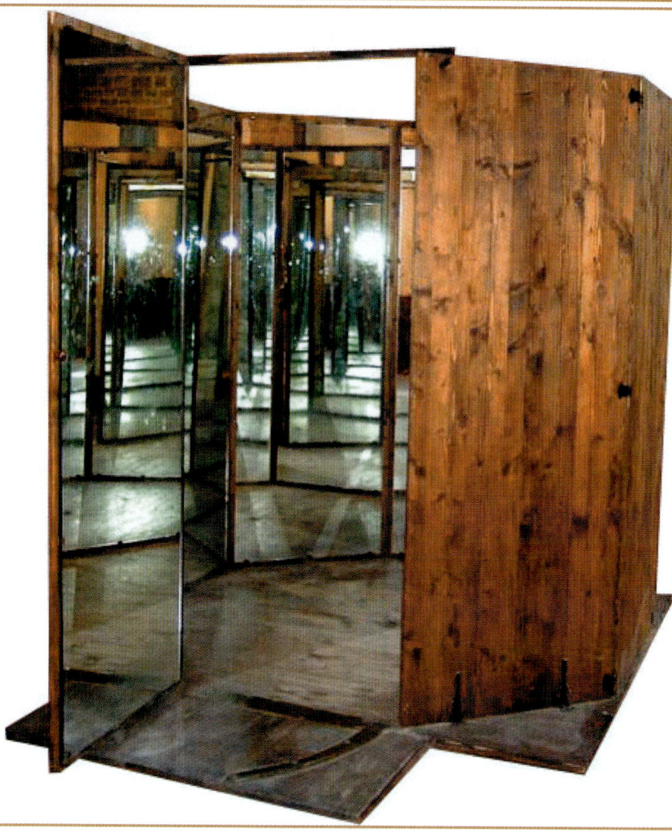

Chamber of Mirrors: Optics were of particular interest to Leonardo. Here he drew, and probably built for a royal entertainment, a chamber of mirrors with eight sides. This device permitted observers to see themselves completely and simultaneously from all sides without moving. A hole, drilled into one panel, also gave the chamber a practical purpose: It afforded observers a 360 degree view of objects inside without having to move around it.

CIRCULAR GEAR

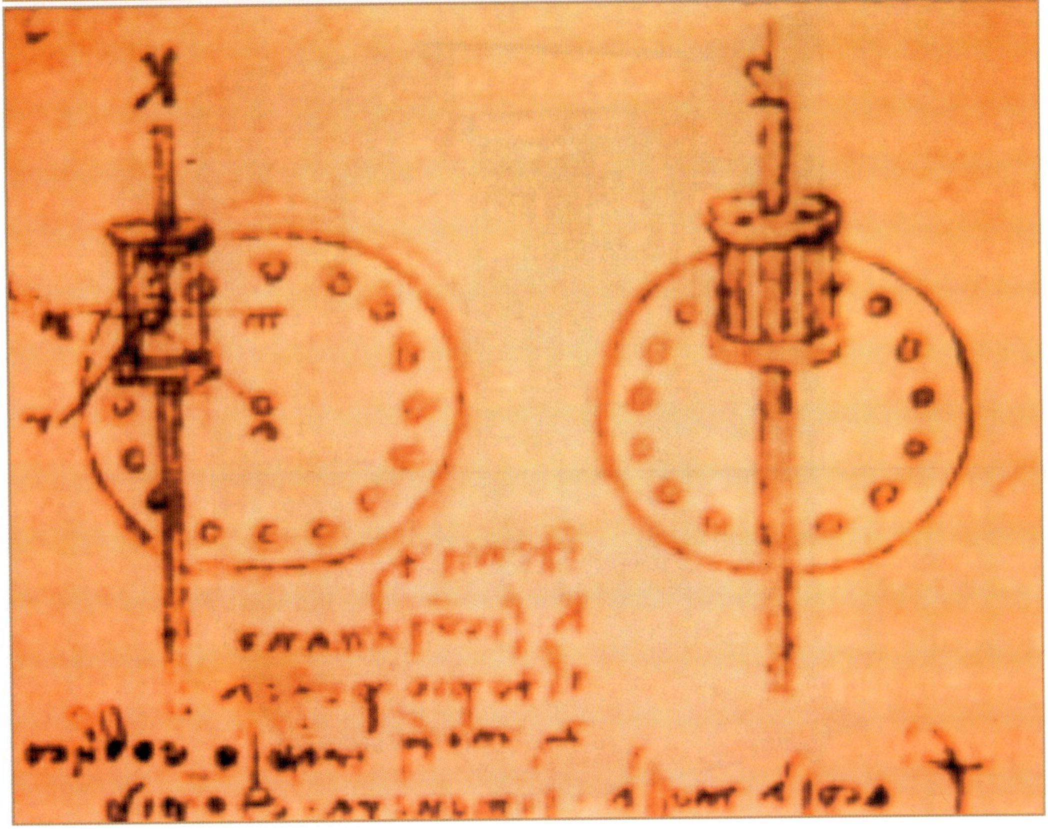

CIRCULAR GEAR
(Ingranaggio a Lanterna)
LV 06
Manoscritto H, f. 38 v

CIRCULAR GEAR

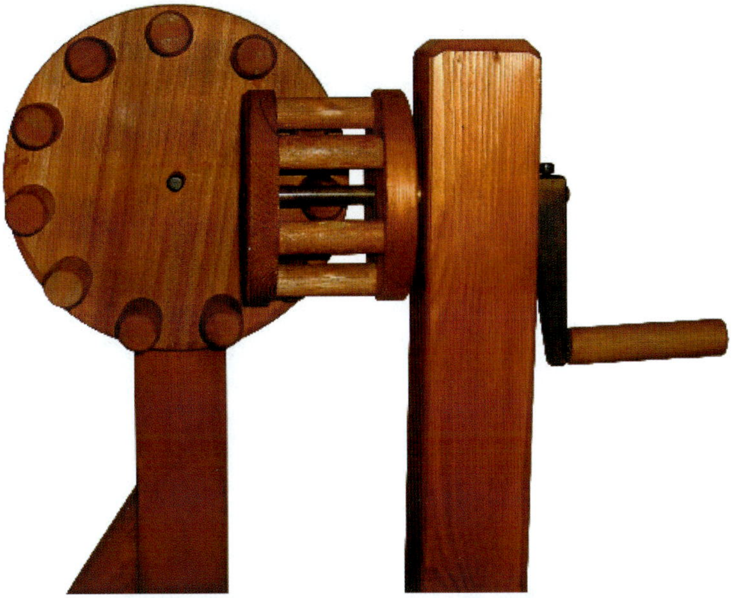

Circular Gear: This system for changing the direction of motion appears often in Leonardo's machines. In order to convert rotary movement from one direction to another, Leonardo frequently used a lantern gear with a cog wheel.

COLUMN HOIST

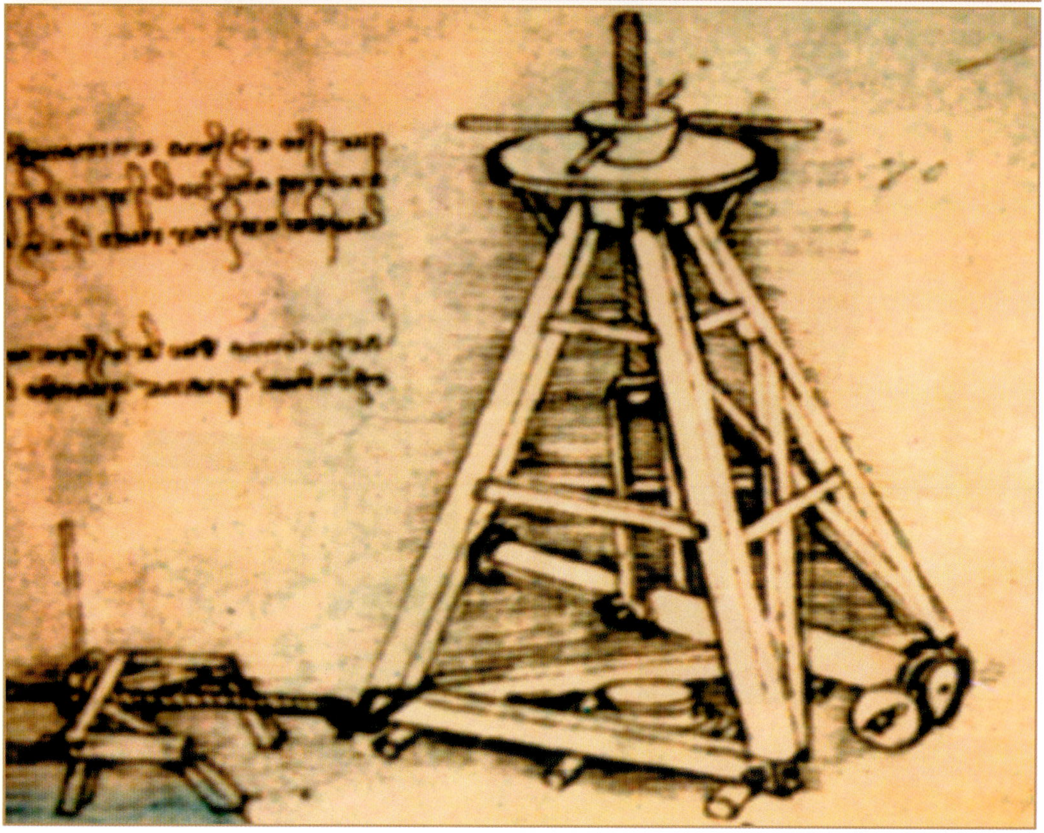

COLUMN HOIST

(Alzacolonne)

LV13

Codex Atlantico f. 49v-a

COLUMN HOIST

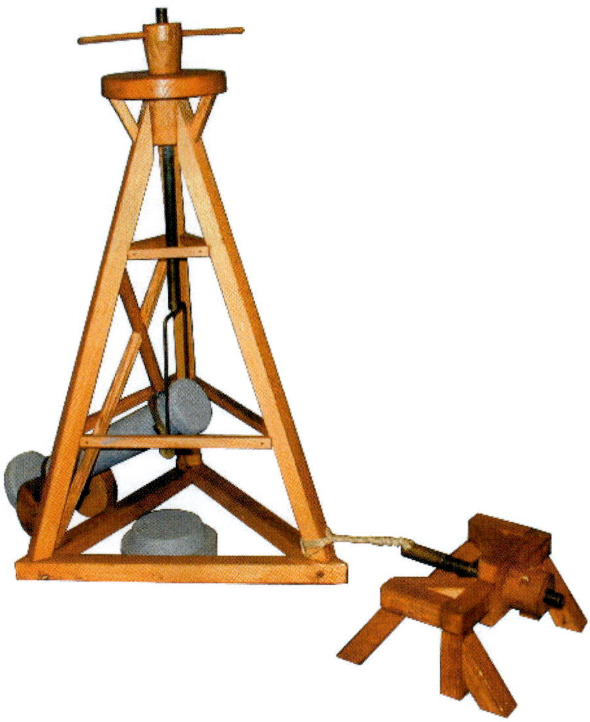

Column Hoist: Leonardo improved on as many older machines as he invented new ones. For example, he designed this machine for raising columns as an improvement on a system devised by other engineers, such as Francesco di Giorgio. In this case, the mechanism is powered by the turn of a crank that sets into motion both a cart on which the column rests and a worm screw mechanism that lifts and holds the other end of the column.

CRANE WITH CENTRAL HOIST

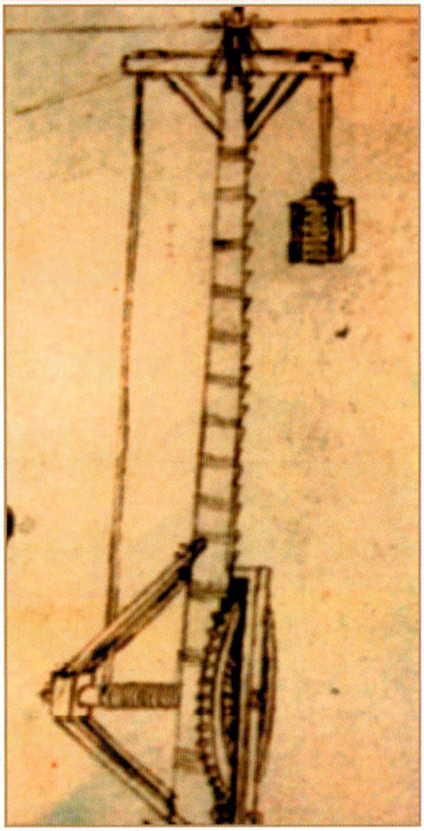

CRANE WITH CENTRAL HOIST
(Gru con Argano Centrale)
LV 63
Codex Atlantico f. 105 bv

CRANE WITH CENTRAL HOIST

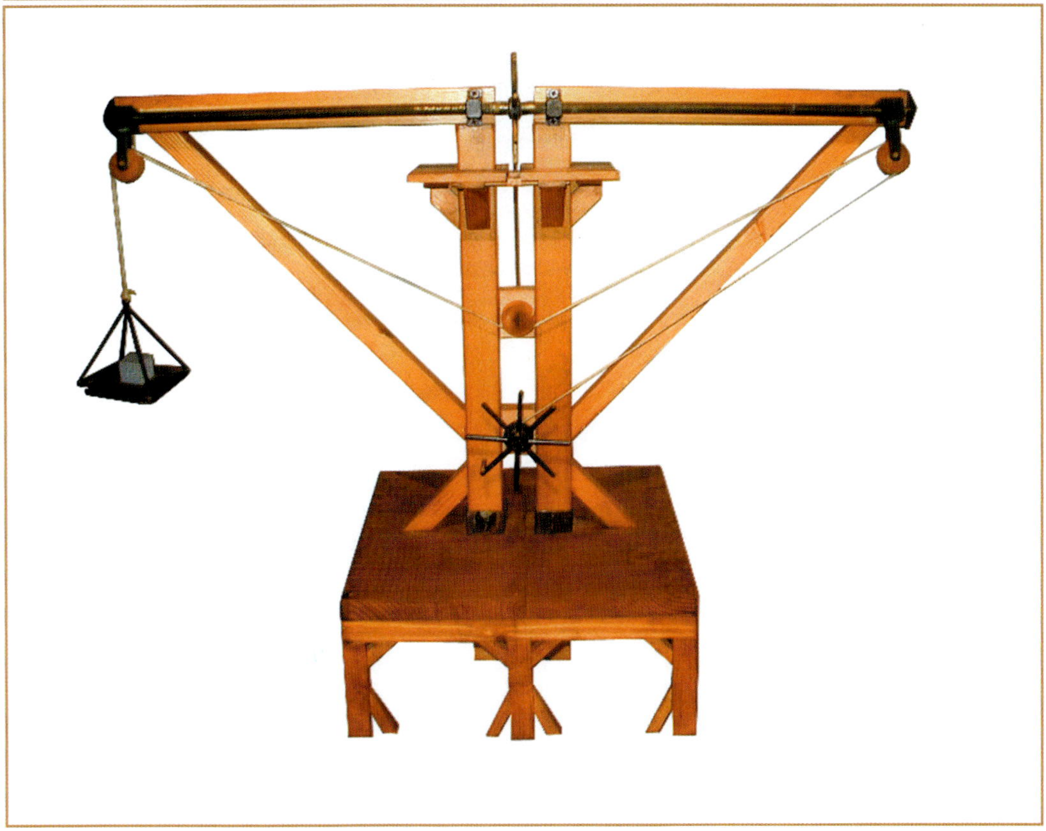

Crane With Central Hoist: This machine is based on one of the cranes Leonardo observed during the period he spent in Florence. He sought to improve these existing cranes by making them more efficient and safer. He concentrated on the cranes designed by the noted 15th-century architect Filippo Brunelleschi.

DIRECTION-CHANGING GEAR

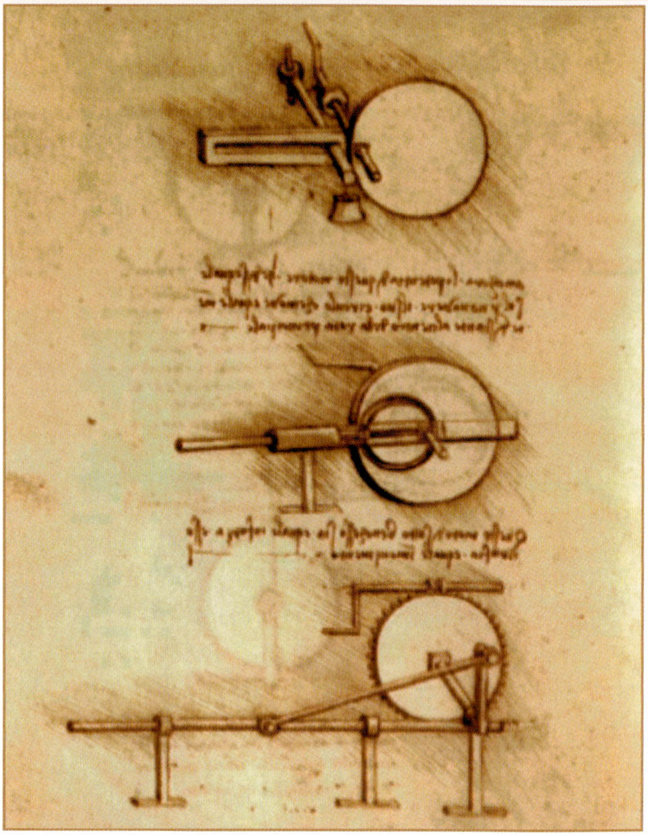

DIRECTION-CHANGING GEAR

(Biella)

LV 40

Codex Madrid I f. 28 v

DIRECTION-CHANGING GEAR

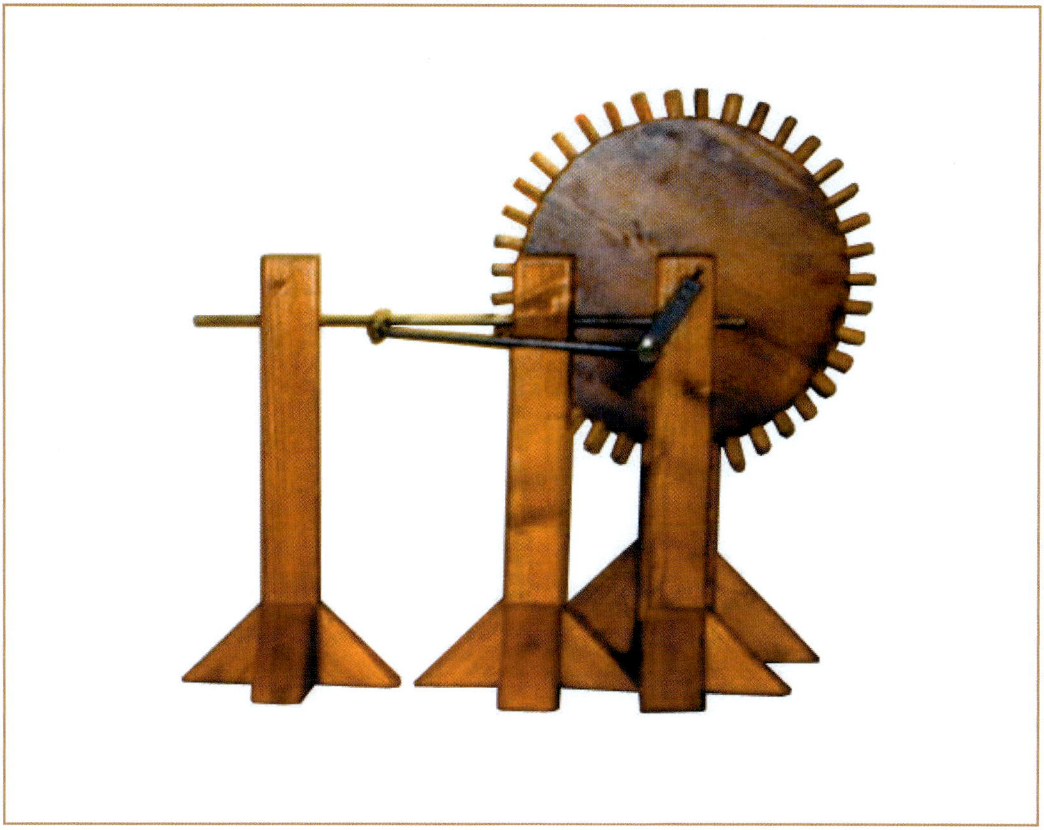

Direction-Changing Gear: Here is yet another very important way Leonardo used gears to transform motion from one mode to another. In this case, the gear wheel transforms its circular motion into the reciprocating, linear motion of the rod.

DIRECTION-CHANGING GEAR — A MODERN USE

Direction-Changing Gear — A Modern Use: Note how the mechanism of a modern windshield wiper follows the same mechanical principle developed by Leonardo. A circular motion is converted to a reciprocating horizontal motion to clear water, dust, debris, or other matter out of the field of vision of the operator of a moving vehicle.

DIRECTION-CHANGING GEAR — A MODERN USE

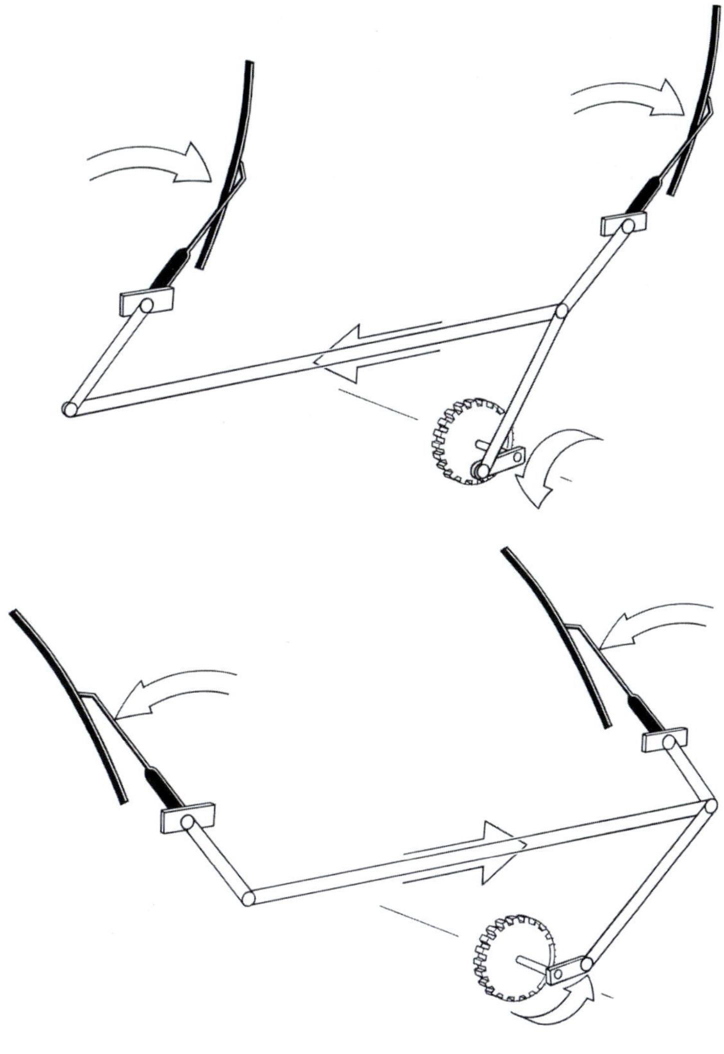

Illustration
Jamie Pfeifer Graphics

EXCAVATOR

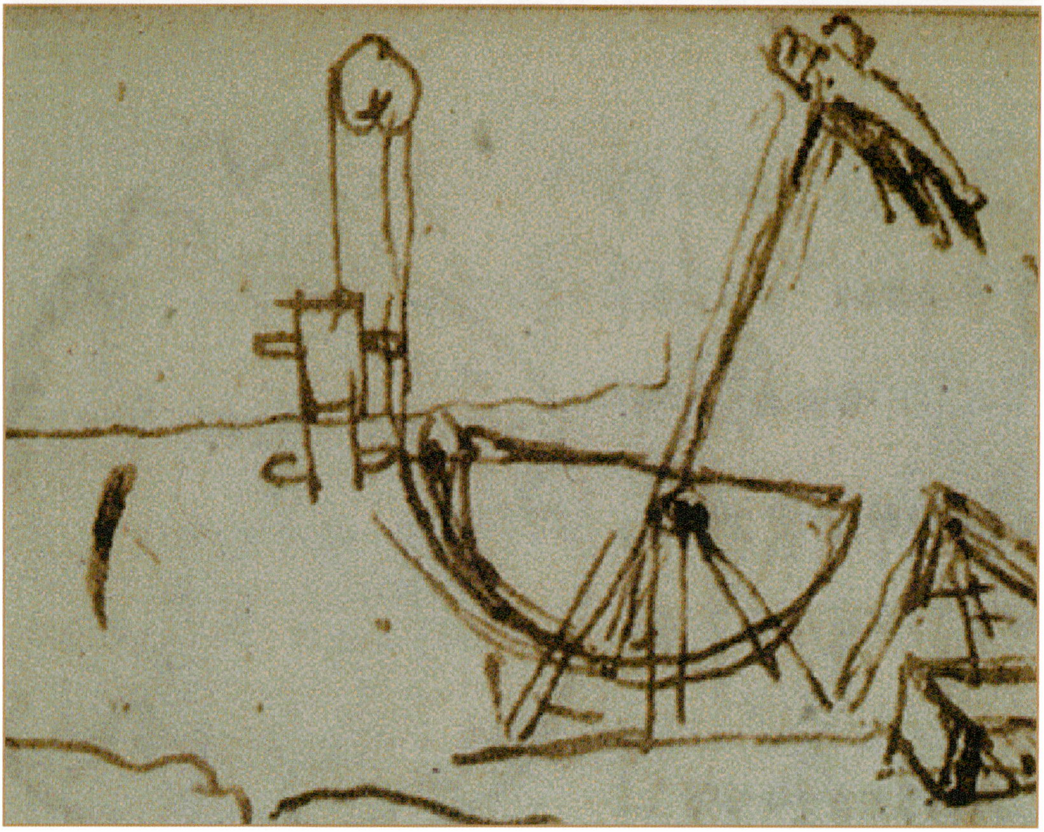

EXCAVATOR
(Escavatrice da Trincea)
LV 17
Manoscritto L f. 76 v

EXCAVATOR

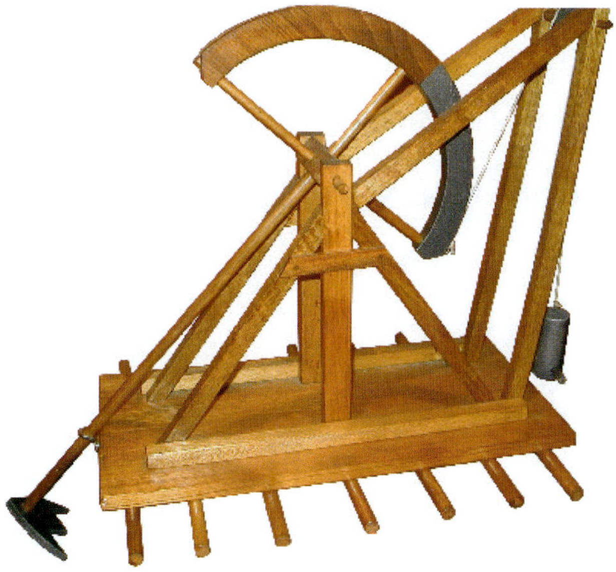

Excavator: This excavation tool was designed by Leonardo to dig trenches. The shovel is operated via the release of the balance weight. The large rocking half-wheel helps lift the shovel and lower the balance weight. The operation is repeated as the shovel digs down into the earth to create a trench. When the proper amount of dirt has been removed, the excavator is moved to a new position to dig a new stretch of earth. Modern backhoes work on the same principle.

FLYWHEEL

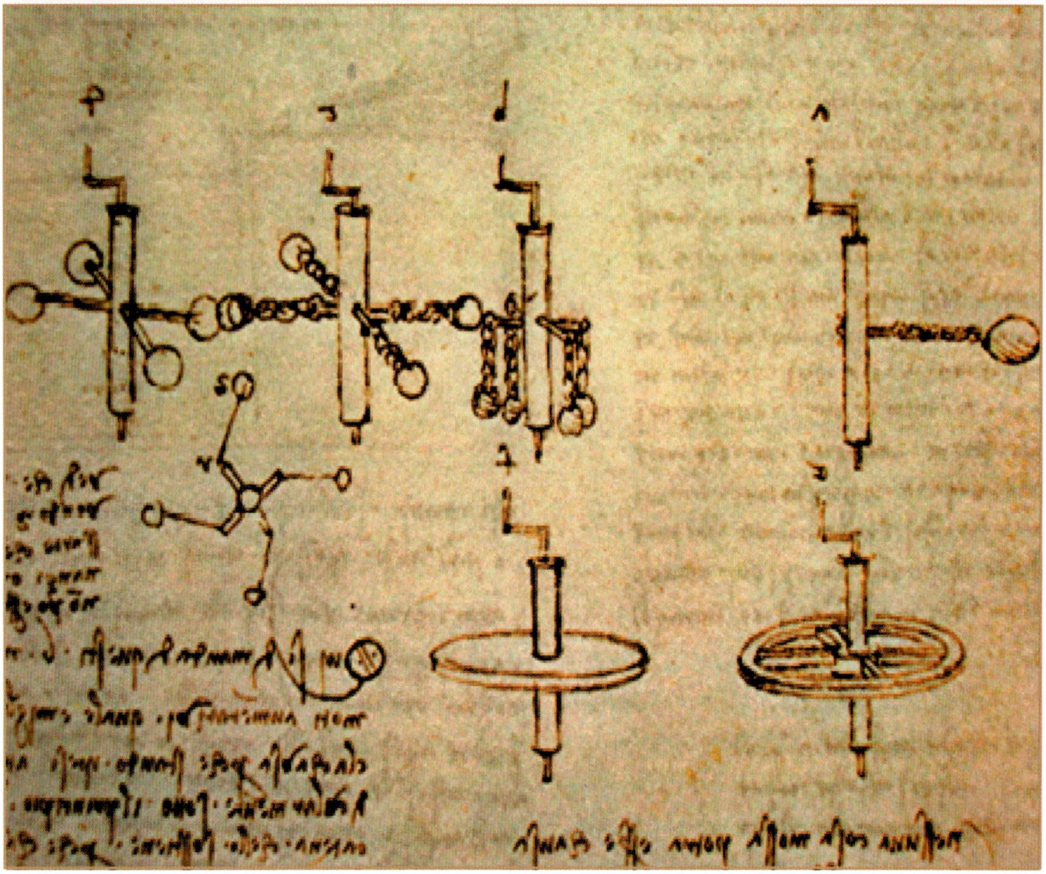

FLYWHEEL
(Volano)
LV 11
Codex Madrid I f. 114 r

FLYWHEEL

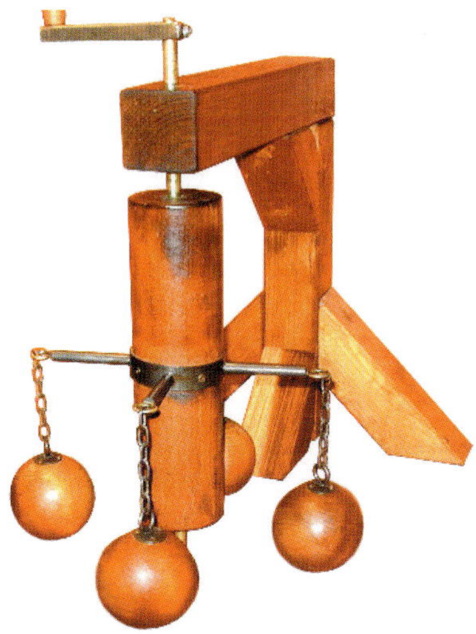

Flywheel: The four balls comprise Leonardo's design of what we now call a flywheel — a heavy device that can store the energy it receives to release it in even amounts to maintain a steady speed. Leonardo worked on ways to allow the flywheel to spin longer yet use less energy to start it turning.

FLYWHEEL — A MODERN USE

Flywheel — A Modern Use: Flywheels are used in many modern mechanisms for precisely the same reason that Leonardo used them: to lessen the need for fresh energy to maintain a specific speed once the desired level has been achieved.

FLYWHEEL — A MODERN USE

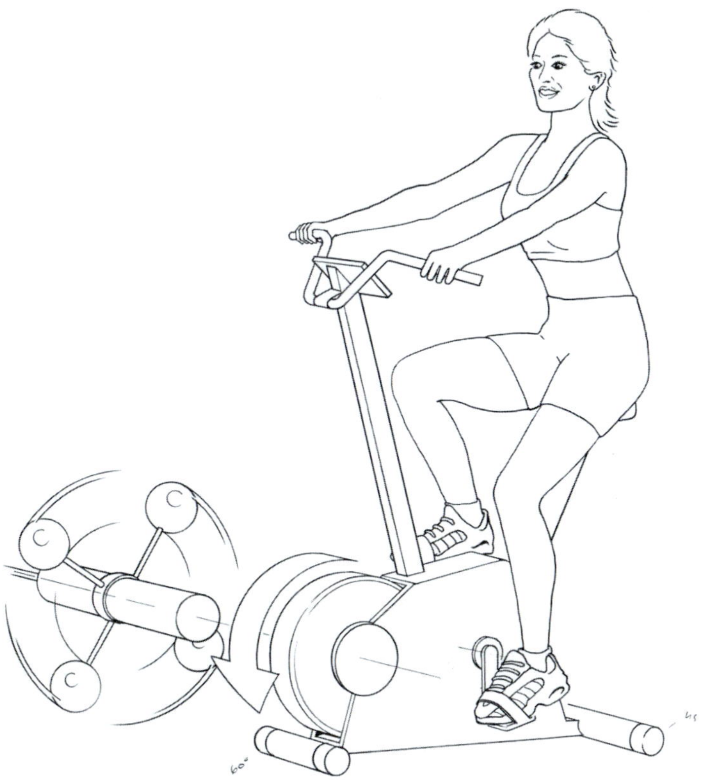

Illustration
Jamie Pfeifer Graphics

FRICTION STUDY

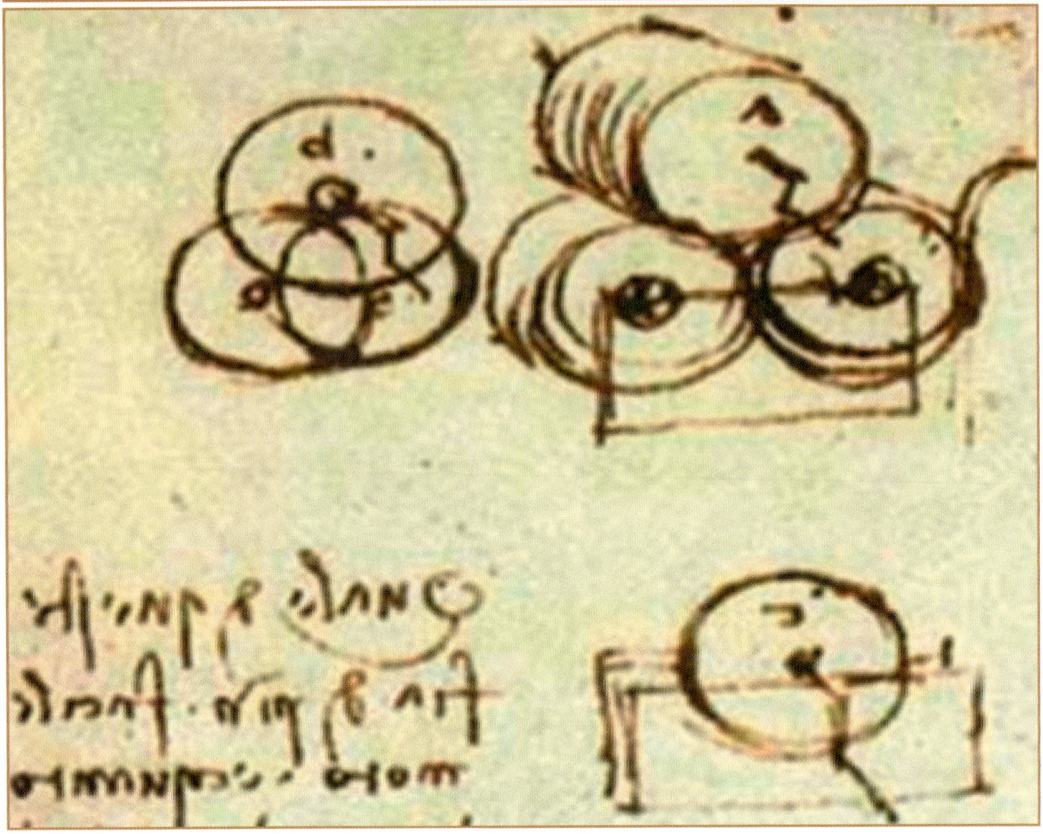

FRICTION STUDY
(Studio per gli Atriti)
LV 70
Manoscritto L f. 11 v

FRICTION STUDY

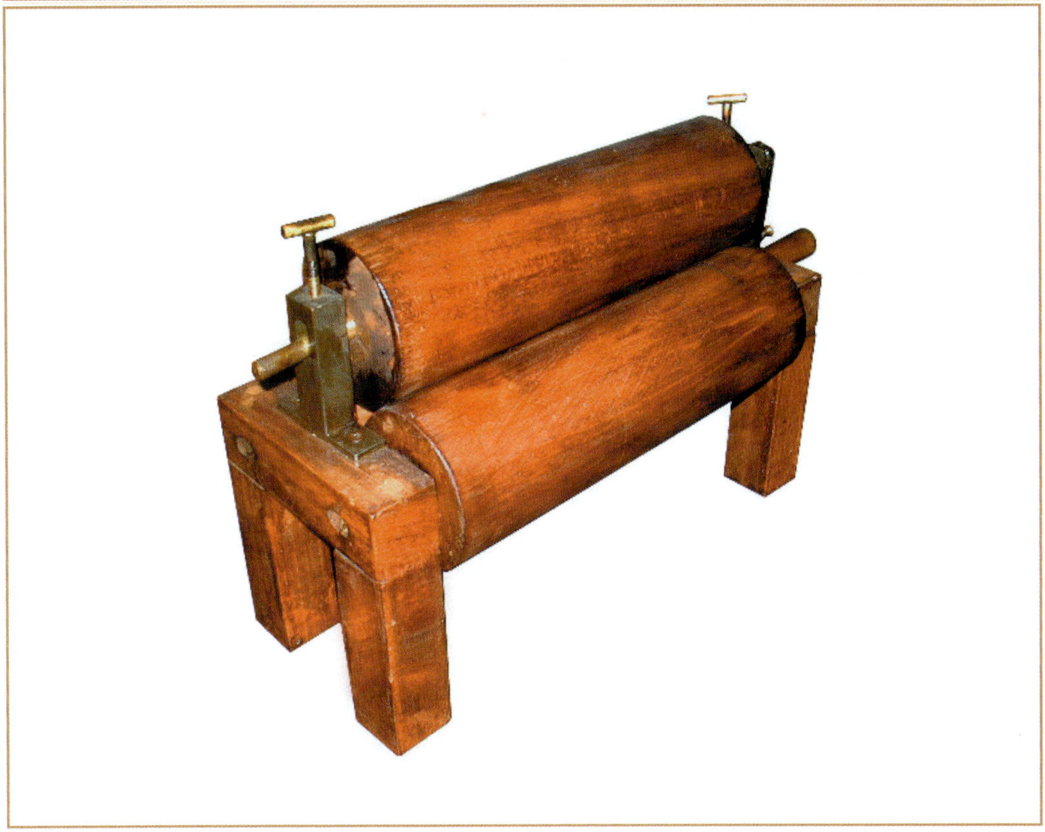

Friction Study: After studying watermills in operation, Leonardo saw that friction was one of the main problems that needed solving. Friction caused extensive wear and tear on the axle attached to the watermill wheel. This idea for an experiment has three drums that turn on themselves to distribute the friction across more surfaces and thus reduce its effect on any one moving part. Modern roller bearings used in heavy machinery — notably on railroad equipment — operate on the same principle.

GEAR STUDY

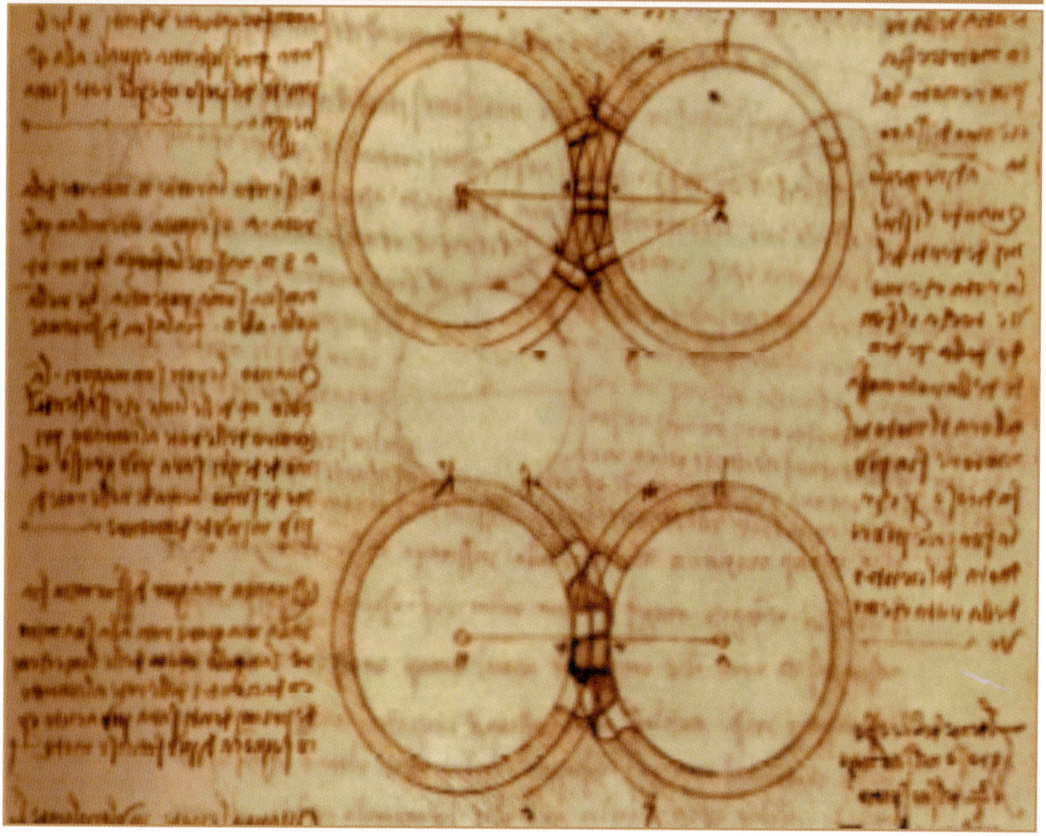

GEAR STUDY
(Studio Ingranaggi)
LV 80
Codex Madrid I f. 5 r

GEAR STUDY

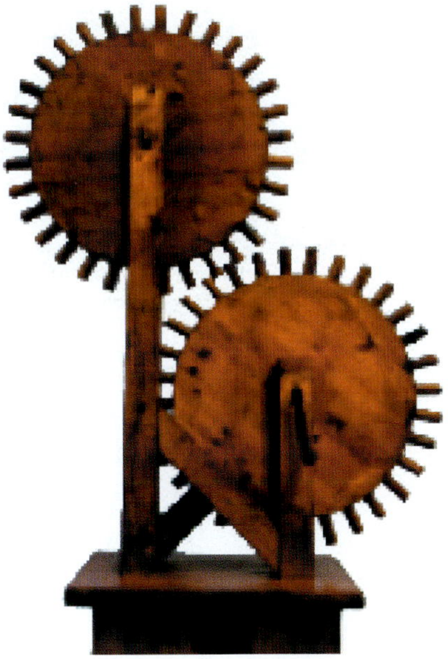

Gear Study: Gears are mechanical devices that transmit motion from one part of a machine to another. They are often used to change the direction of a motion — from circular to horizontal, for example — or the speed of movement. Turning the wheel at the bottom and the interlocking sprockets move the wheel at the top. Leonardo explains that the "power of the motion made with the help of toothed wheels" varies according to whether the teeth of one wheel "mesh with the bases, the middle, or the tips of the teeth of the opposite wheel." His experiments helped him determine that maximum power is obtained with a tip-to-tip combination.

HEAVY-LIFTING MECHANISM

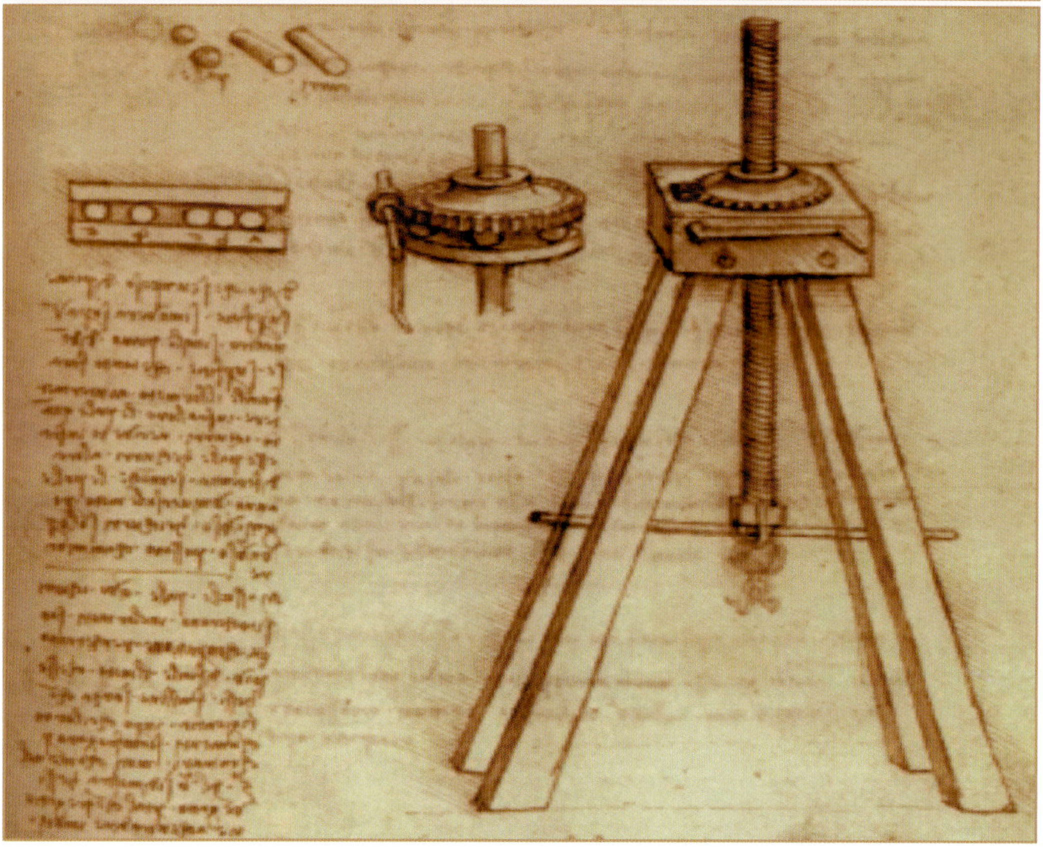

HEAVY-LIFTING MECHANISM
(Binda a Vite)
LV 82 / 83
Codex Madrid I f. 26 r

HEAVY-LIFTING MECHANISM

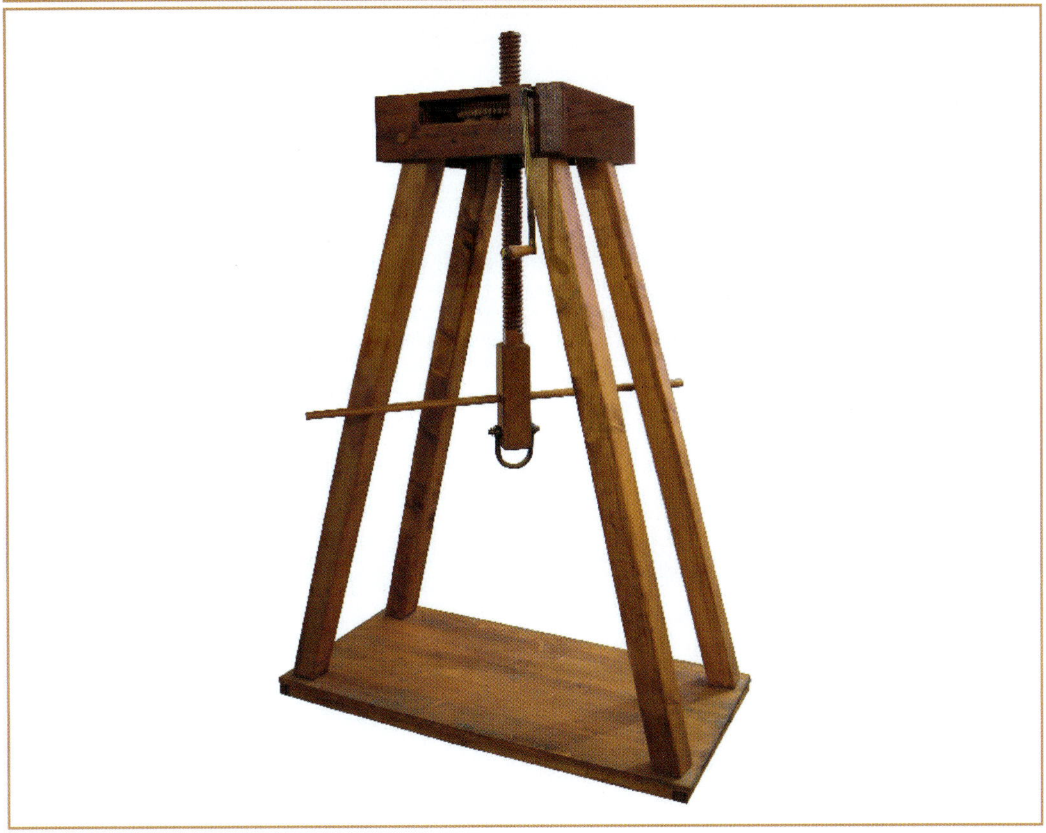

Heavy-Lifting Mechanism: From his earliest days in Florence as a young apprentice artisan, Leonardo was fascinated with mechanics. He began by drawing machines employed by architects of the time, particularly lifting devises. He then systematically studied ways of improving them. This machine to lift extremely heavy objects has innovations in three areas: the crank handle on the center pole makes it very easy to use, the ball bearings at the top reduce friction, and the worm screw makes it both safe and powerful.

HYDRAULIC DRILL

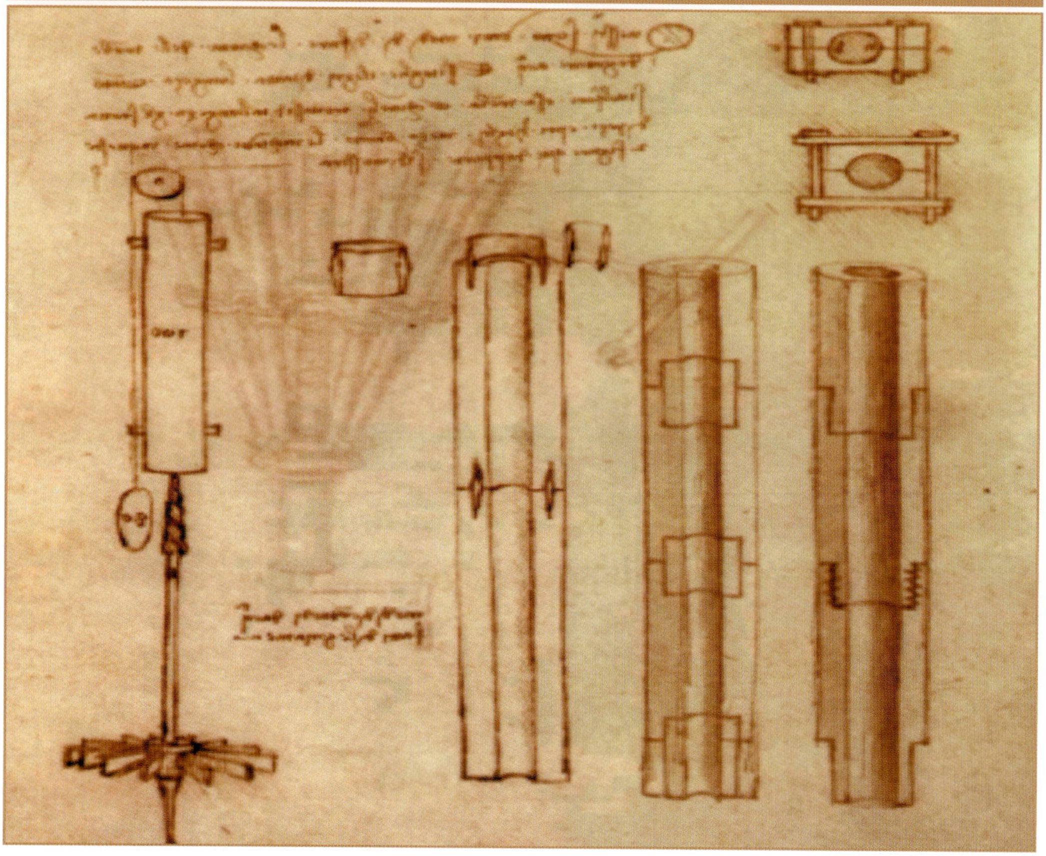

HYDRAULIC DRILL
(Trivella Idraulica per Tronchi)
LV 81
Codex Madrid I f. 25 v

HYDRAULIC DRILL

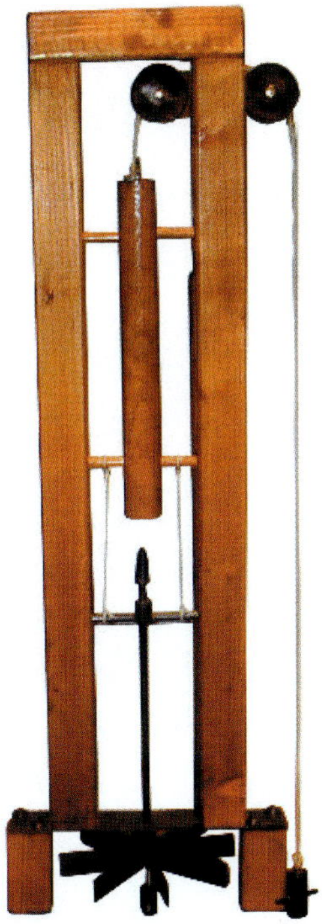

Hydraulic Drill: The purpose of this machine was to drill a hole the length of a log — a 15th century technology designed to create pipes capable of carrying water and waste. The drill is powered by a horizontal hydraulic wheel and driven by the force of a river. Leonardo also designed the connections to permit several such pipes to be fitted together. Wooden pipes were used well into the 19th century.

HYDRAULIC SAW

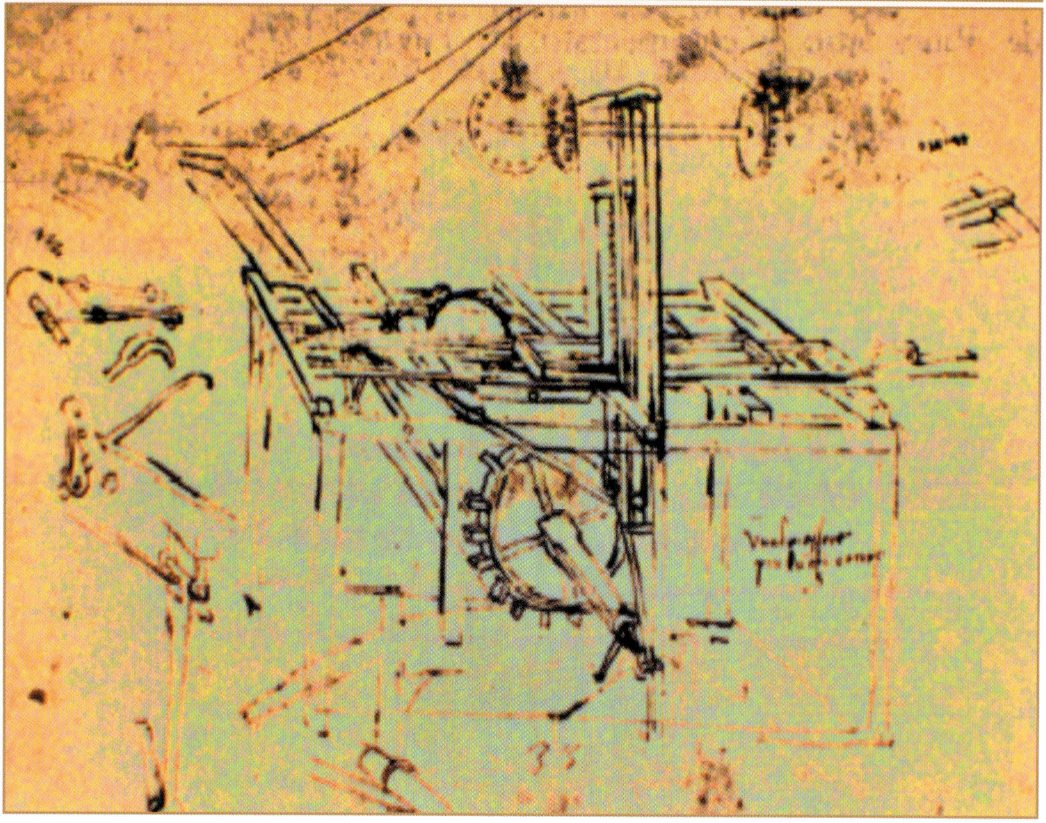

HYDRAULIC SAW
(Sega Idraulica)
LV 69
Codex Atlantico f. 1078 a-r

HYDRAULIC SAW

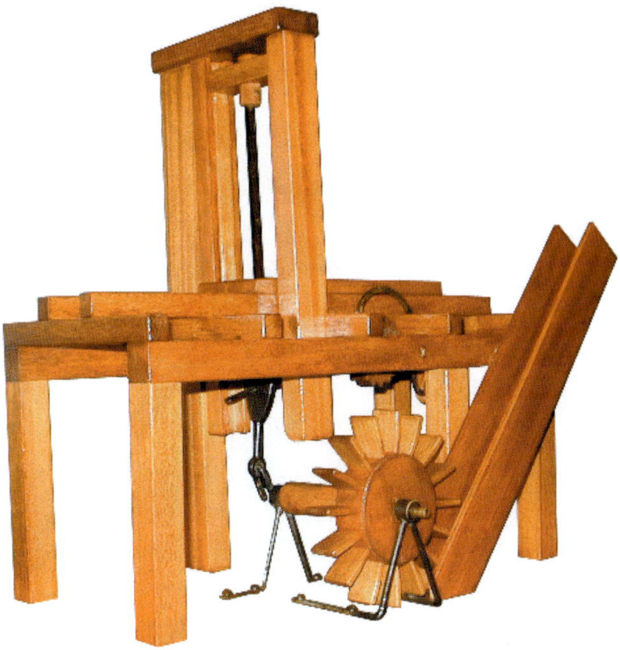

Hydraulic Saw: While Leonardo often was intent on making improvements to machines then in wide use, his automated hydraulic saw was a first. It transmits the power of the river through a series of gears to the alternating motion of the blade. The blade is thus able to cut through a tree trunk moving on a carriage between the water wheel and the blade. This type of saw can still be found in use in some Alpine regions today.

LONG POLE HOIST

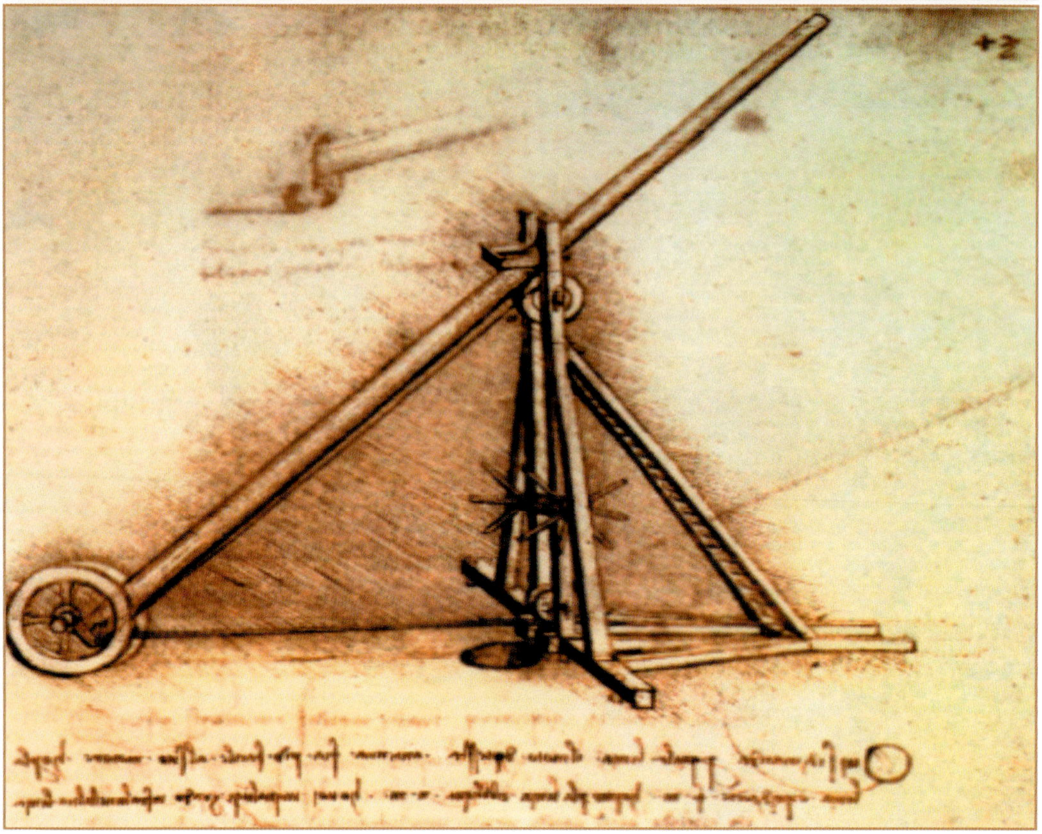

LONG POLE HOIST
(Gru per Innalzare Pertiche)
LV 62
Codex Madrid I f. 43 r

LONG POLE HOIST

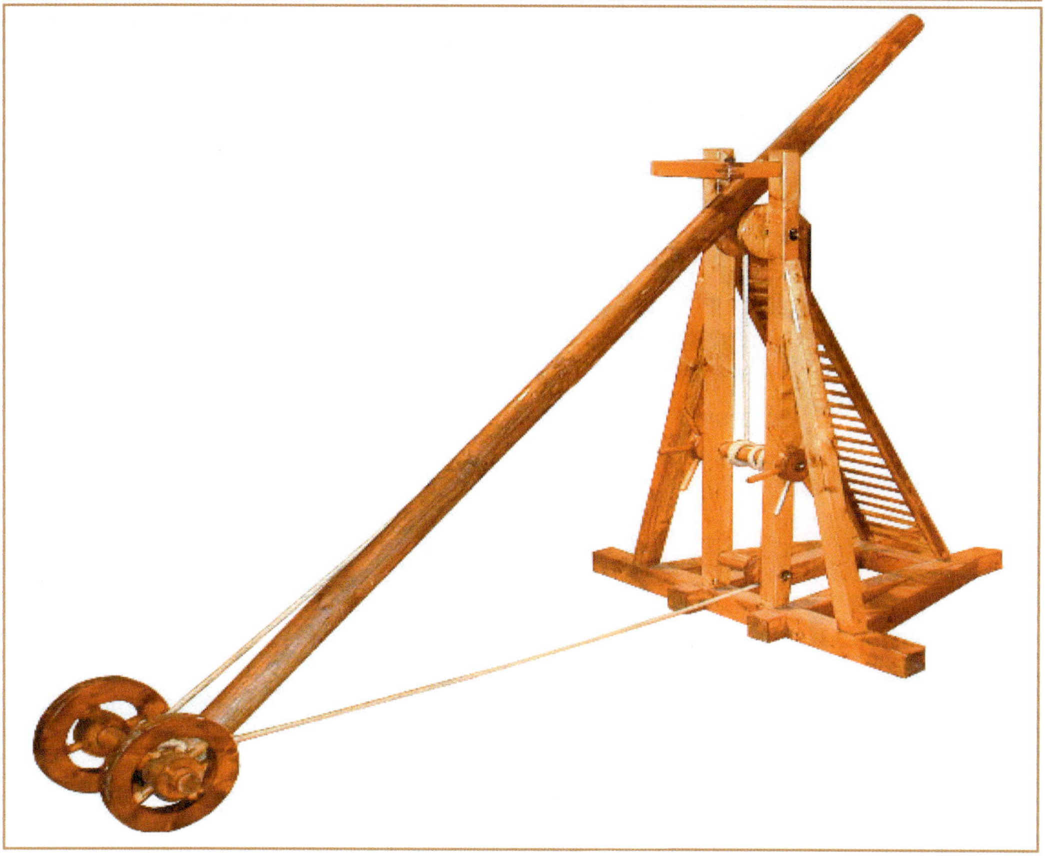

Long Pole Hoist: A rope connected to a winch pulls the pole horizontally from the base, considerably diminishing the need for human effort. The wheels also help by reducing friction. The drawing and Leonardo's comment can be connected to his interest in geometry and, in this particular case, the triangle. (See also his Column Hoist for another device he developed to raise long, heavy objects.)

NIGHT CLOCK

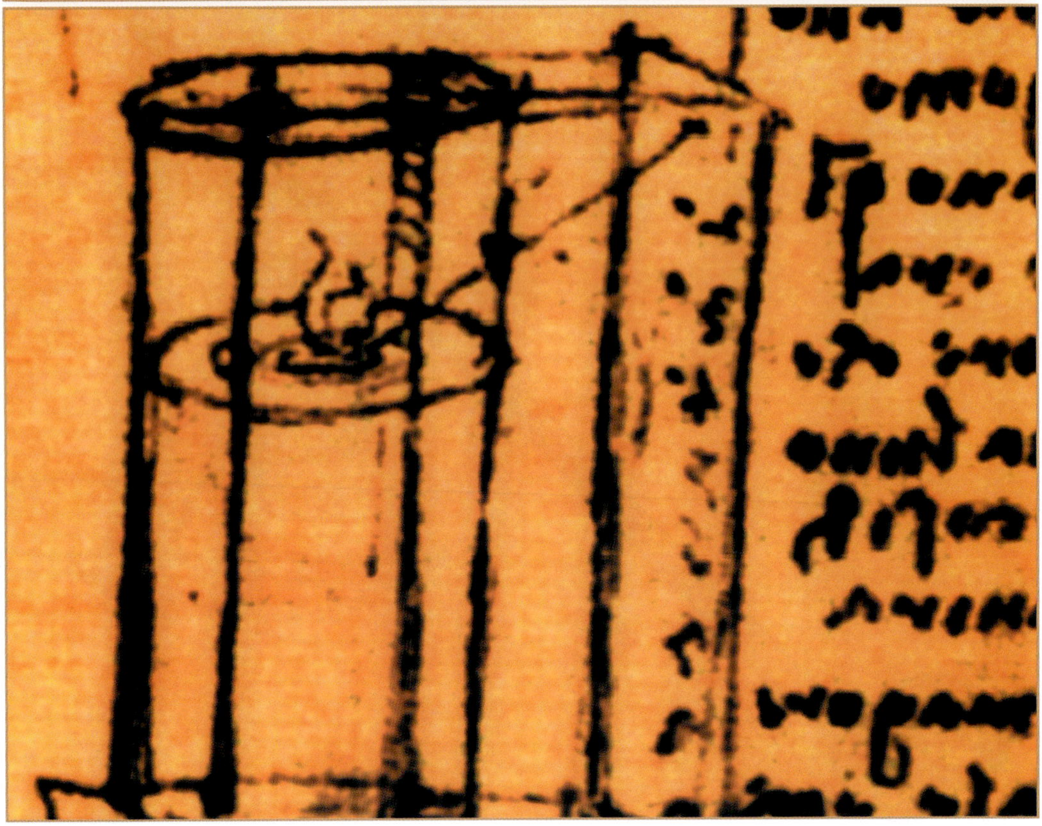

NIGHT CLOCK
(Orologio a candela)
LV78
Codex Atlantico f 394 v

NIGHT CLOCK

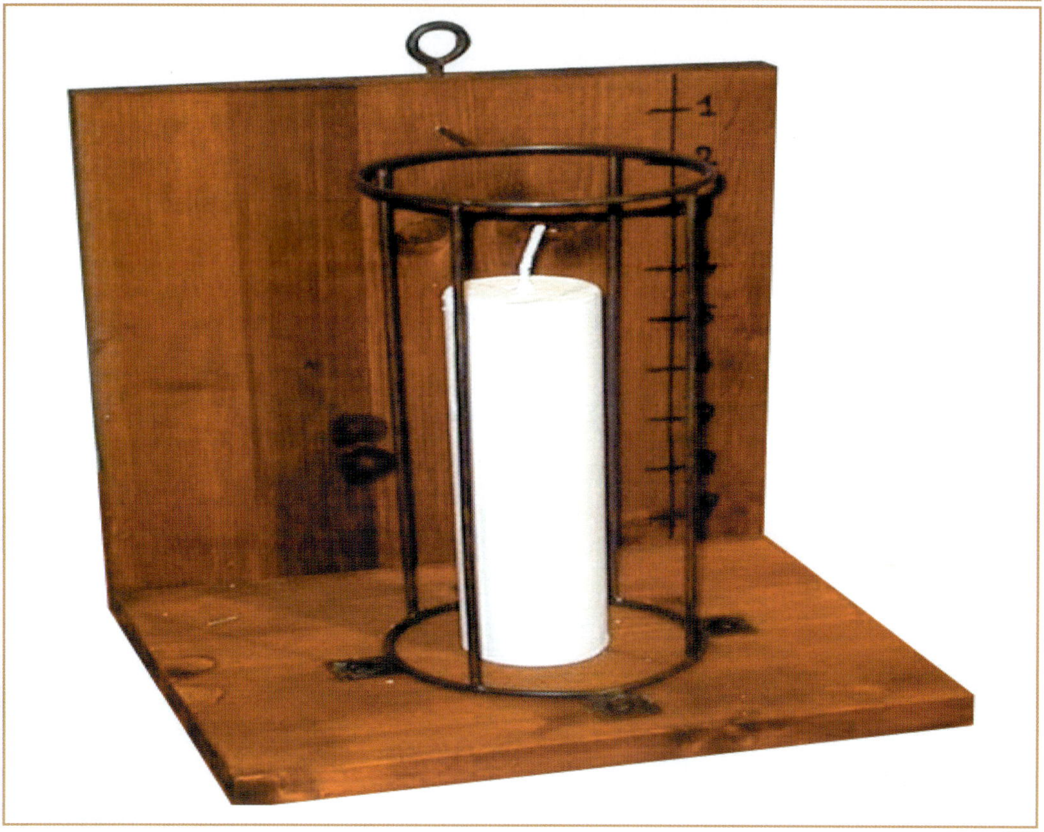

Night Clock: Telling time at night was a problem until Leonardo combined a candle of a specific diameter with a precise scale to calculate the passage of time during the period between dusk and dawn. The patience and persistence he exhibited to find the right wax, right diameter, right type of wick, and an understanding of the impact of temperature and moisture on burning rates, revealed a pattern that would mark all of Leonardo's efforts as a designer and engineer: if it wasn't perfect, it wasn't finished.

PERPETUAL MOTION STUDY

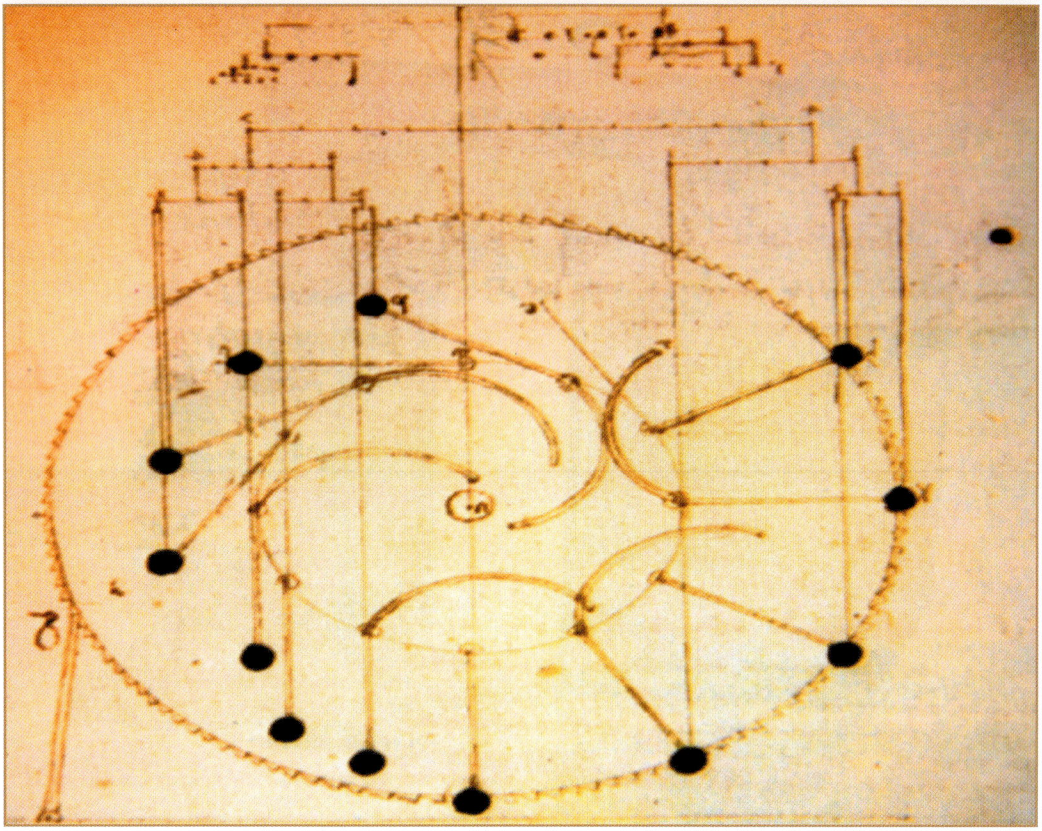

PERPETUAL MOTION STUDY
(Studio del Moto Perpetuo)
LV 47
Codex Forster II f. 90 v

PERPETUAL MOTION STUDY

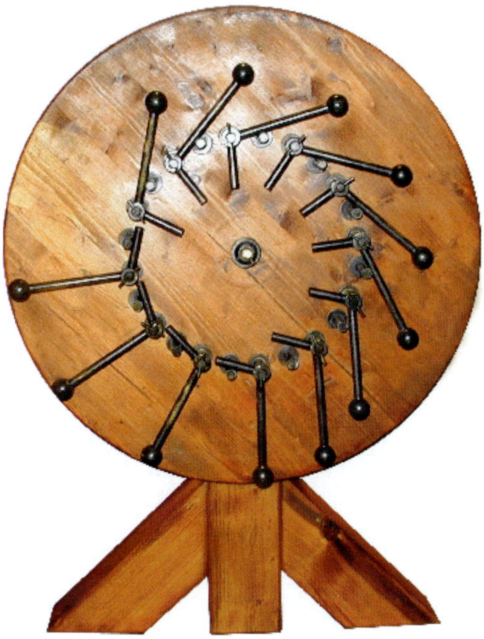

Perpetual Motion Study: During Leonardo's time, there was much discussion about creating a machine that could operate continuously without any subsequent input of power after being started. Leonardo used drawings and comments to assert the impossibility of achieving what we now call perpetual motion. The example here was built from sticks with weights attached to the ends to demonstrate that no matter the design, some form of fresh external force needs to be applied to keep a machine running.

PULLEYS: LARGE

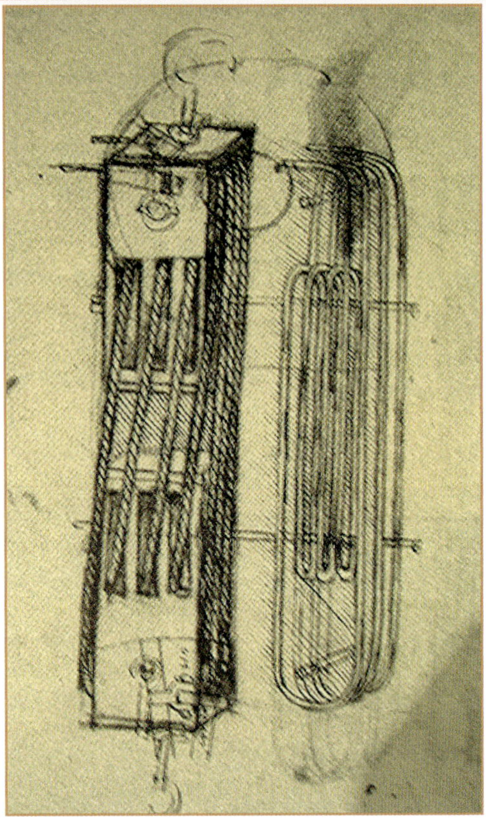

LARGE PULLEYS
(Pesi con Carrucole)
LV 82
Codex Atlantico f. 396 r

PULLEYS: LARGE

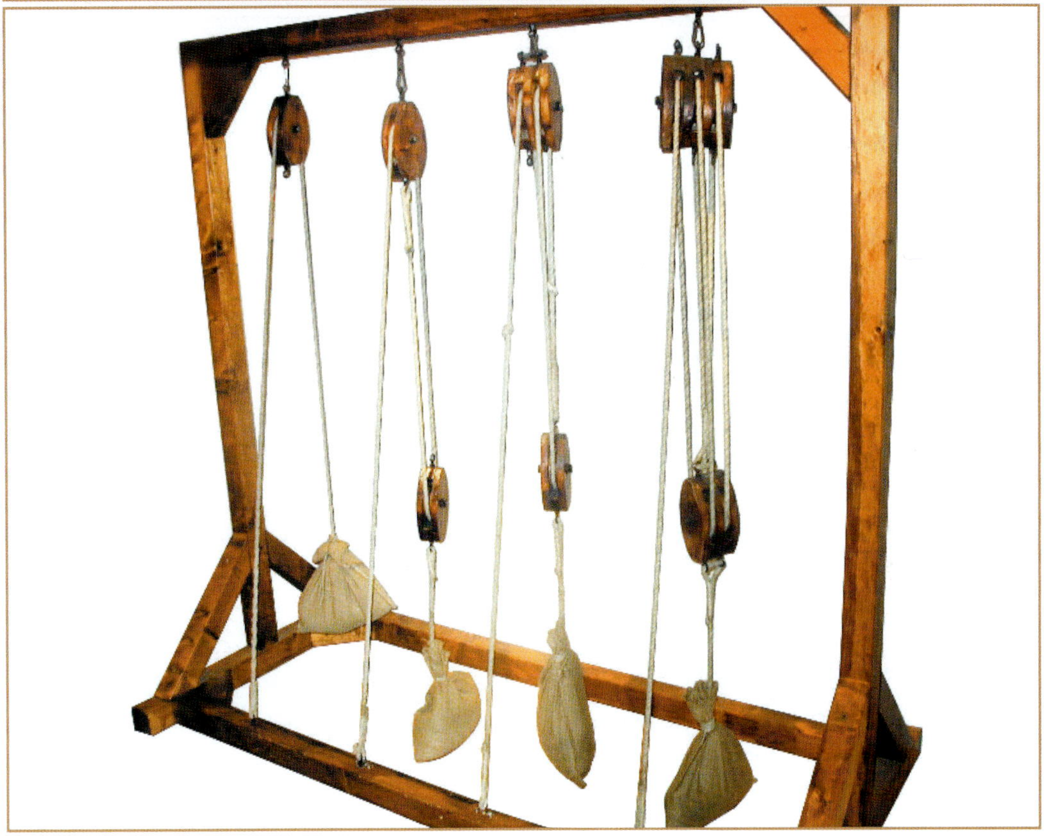

Large Pulleys: This study shows the effect of pulleys on the movement of weighty objects. Each of the sand bags weighs the same. When the rope on the far left is pulled down, the single pulley raises the bag, but when two pulleys are employed, the bag becomes easier to raise even though it doesn't move as far with the same downward pull. As a result, Leonardo could demonstrate the tradeoff between speed and effort.

PULLEYS: LINKED

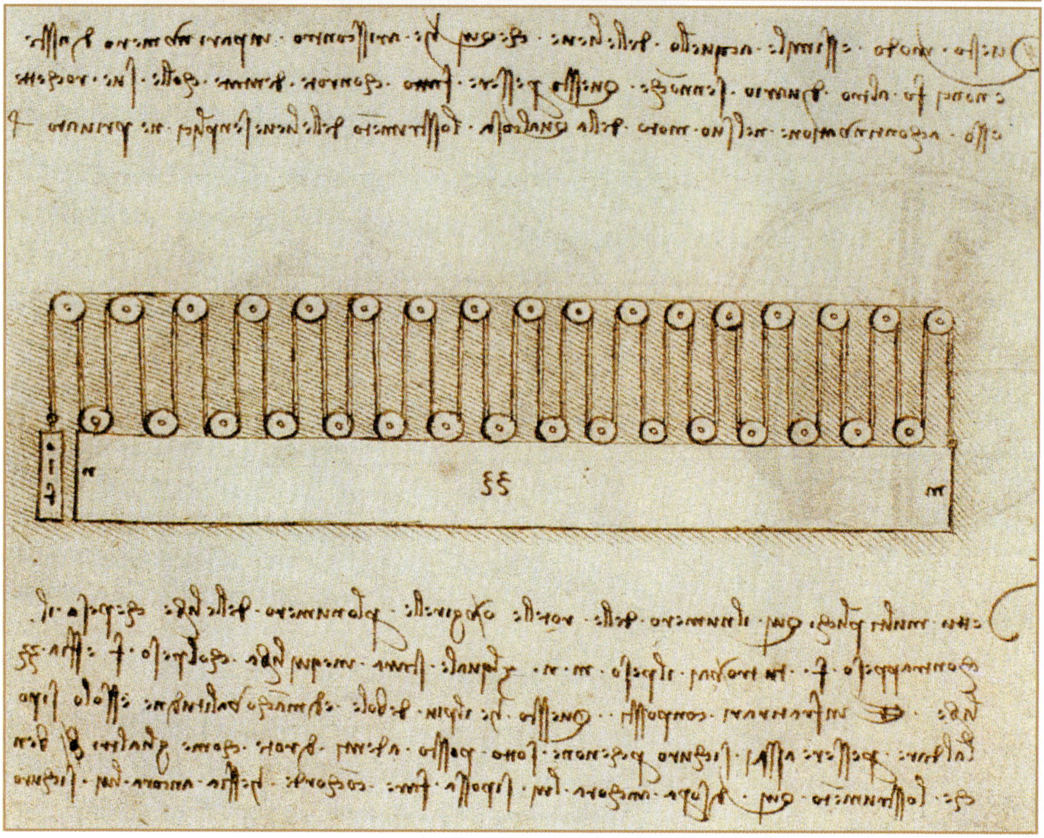

LINKED PULLEYS
(Carrucole)
LV 29
Codex Madrid I f. 36 v

PULLEYS: LINKED

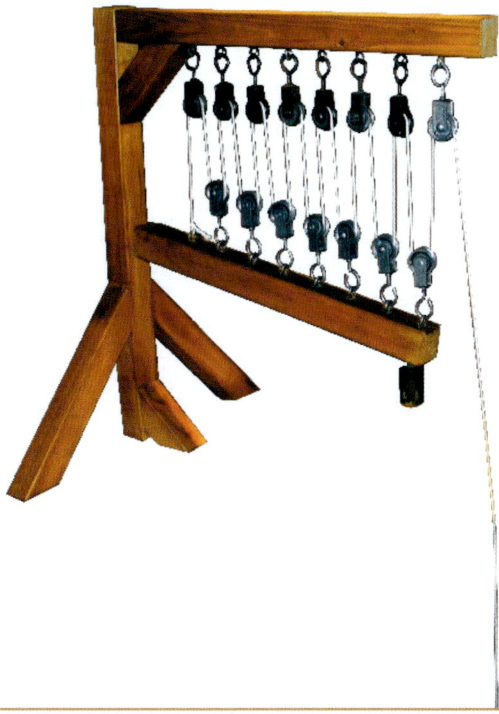

Linked Pulleys: While in Milan, Leonardo conducted extensive studies on ways to reduce the effort necessary to lift heavy objects. He designed various types of pulley mechanisms, some simple and some extremely complex. In this study of linked pulleys, Leonardo demonstrates the way one person can lift long and heavy objects with greatly reduced effort. This is achieved by distributing the weight of the object over a number of equally spaced pulleys.

ROBOT

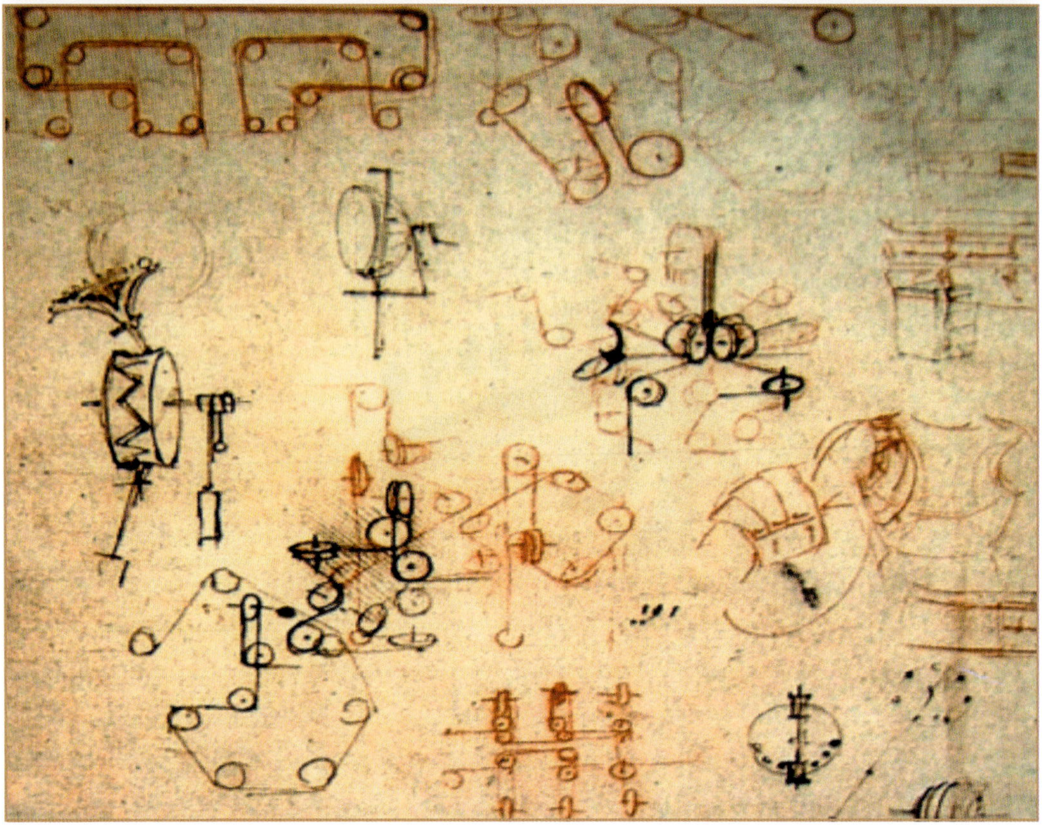

ROBOT
(Robot)
LV 39
Codex Atlantico f. 579 r

ROBOT

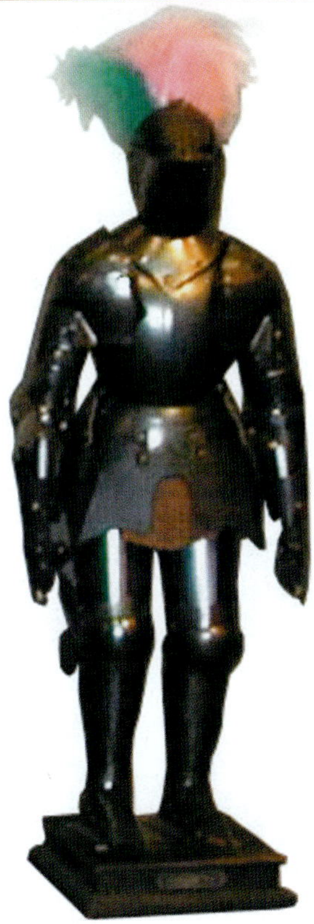

Robot: This is one of Leonardo's least known designs, probably developed for the entertainment of royal guests and informed by his extensive studies of the human body. The robot was designed to open and close its jaw (which was anatomically correct), sit up, wave its arms, and move its head with the use of hidden ropes and pulleys. Because it was intended to be seen from a distance and to inspire wonder, the robot would probably have been larger than the largest man of the day.

SPOTLIGHT

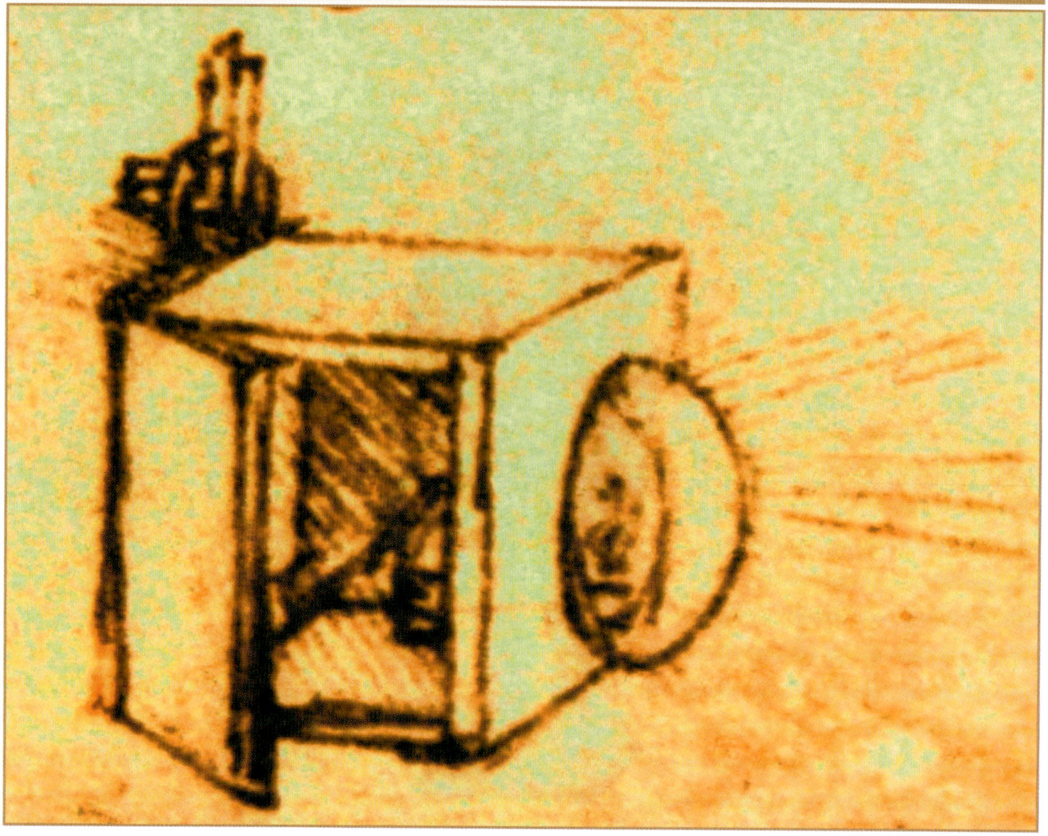

SPOTLIGHT
Reflector / Projector (Proiettore)
LV 71
Codex Atlantico f. 34 r

SPOTLIGHT

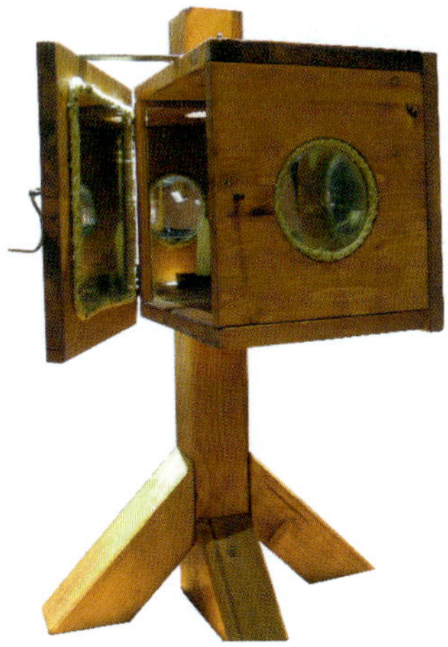

Spotlight: This device is a simple box with a large glass lens on one side and a candle with mirrors inside to be able to make, in Leonardo's words, "an intense and wide light." The principle is the same as that used in floodlights and was undoubtedly intended for use in the theatrical productions Leonardo staged for his royal patrons.

SPRING

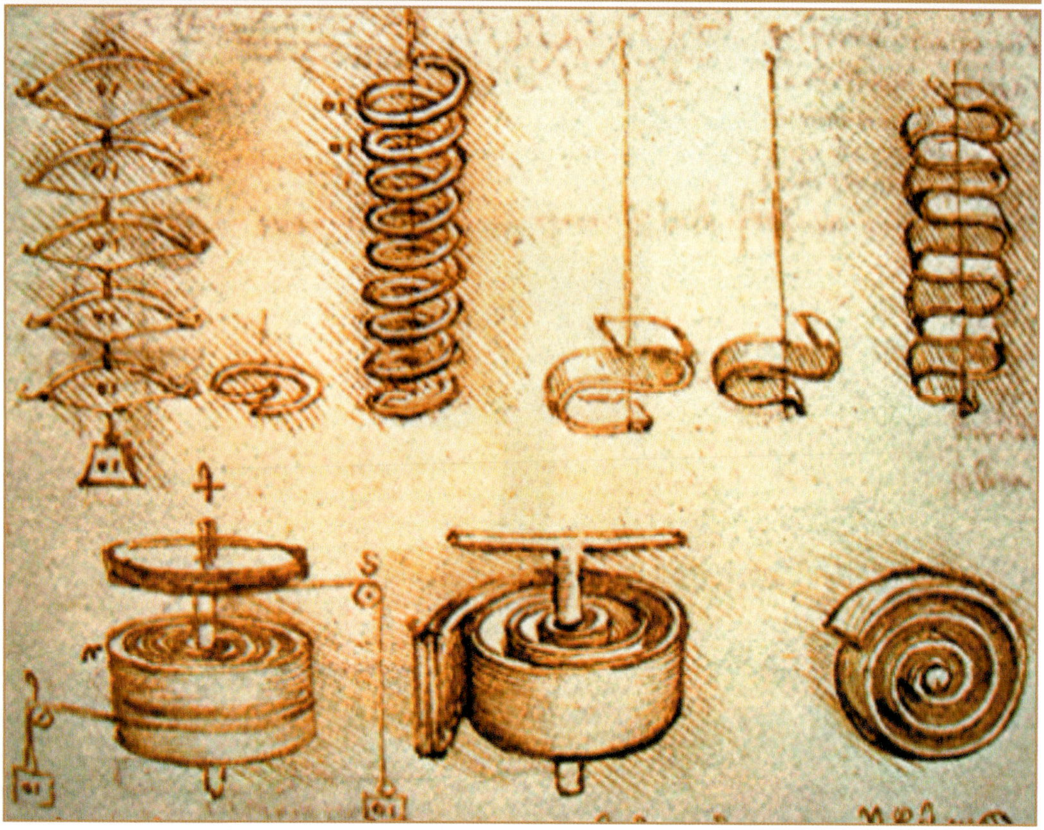

SPRING
(Molla)
LV 8
Codex Madrid I f. 85 r

SPRING

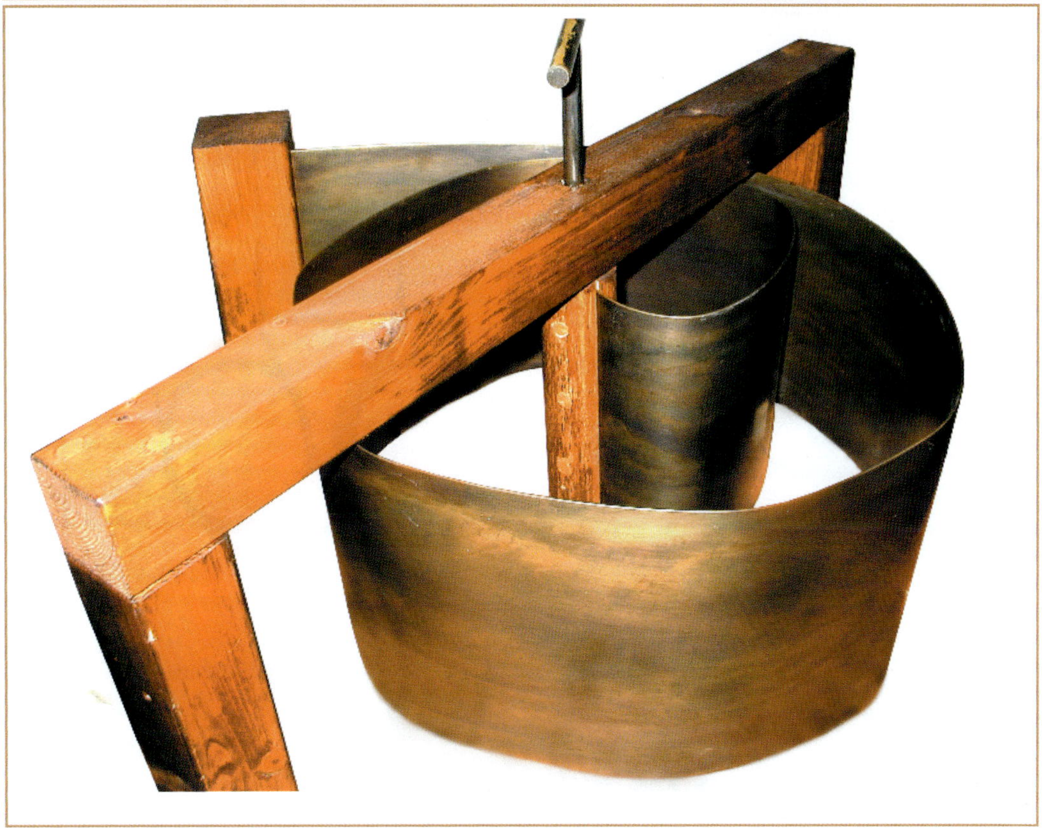

Spring: One of Leonardo's great preoccupations involved the question of how to capture and store energy for later use. This spring is such a device. It was employed to release energy to start machines that could then use wind or water as the continuing source of energy. Leonardo drew many examples of springs during his study of mechanical devices. His efforts were impeded by the tensile-strength of the materials available in his time.

THREE BALL BEARINGS

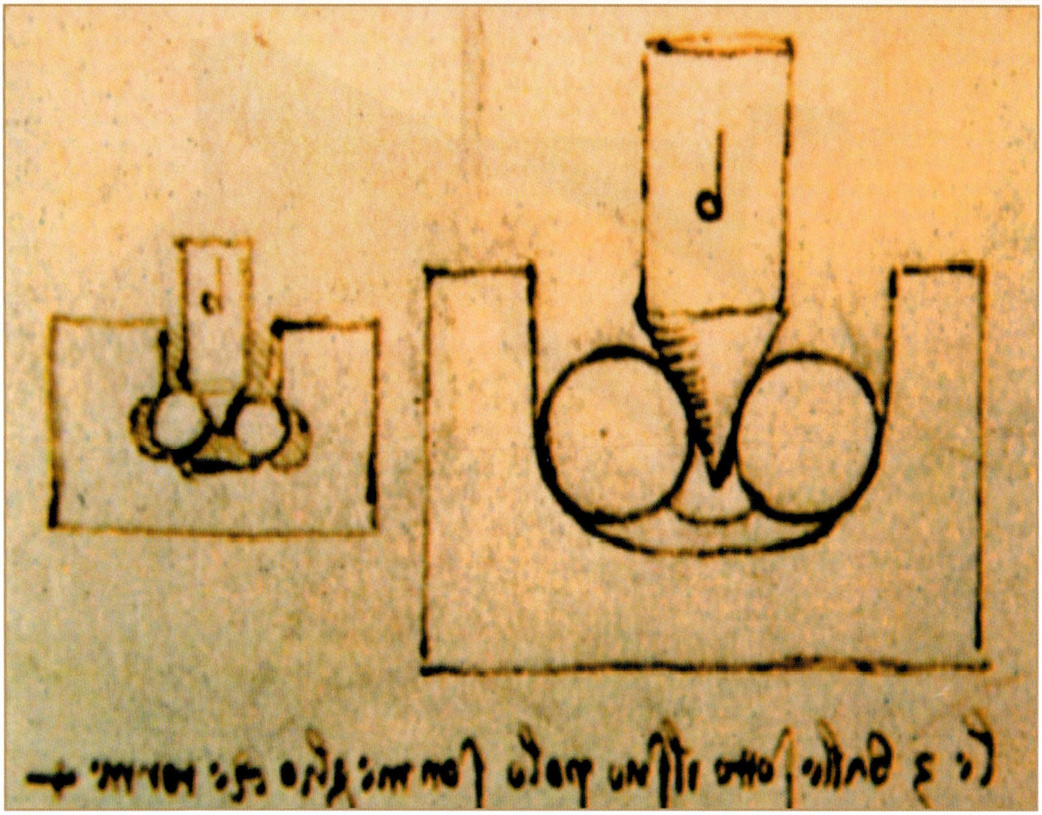

THREE BALL BEARINGS
(Cuscinetto a Tre Sfere)
LV 34 / LV35
Codex Madrid I f. 101 v

THREE BALL BEARINGS

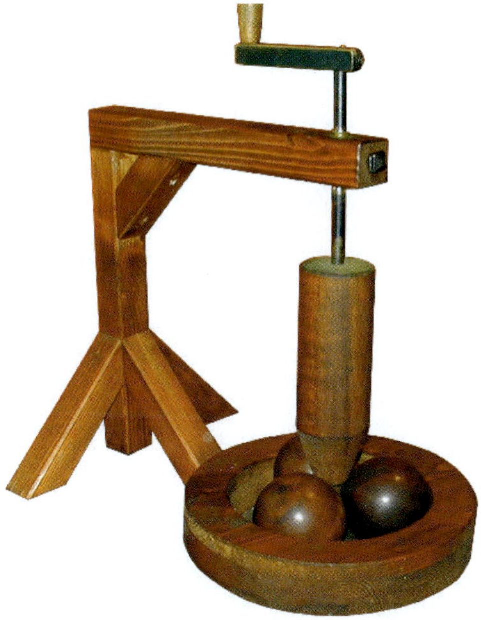

Three Ball Bearings: In order to limit the friction that develops when two moving surfaces rub together — which causes wear and eventual damage to the parts — Leonardo saw a way to separate the parts with a series of balls. Leonardo improved upon the principle by devising a ring-shaped track along which the balls could move without coming into contact with each other. (See also "Ball Bearings" in this section.) The application of this principle is still widely used today.

TONGS

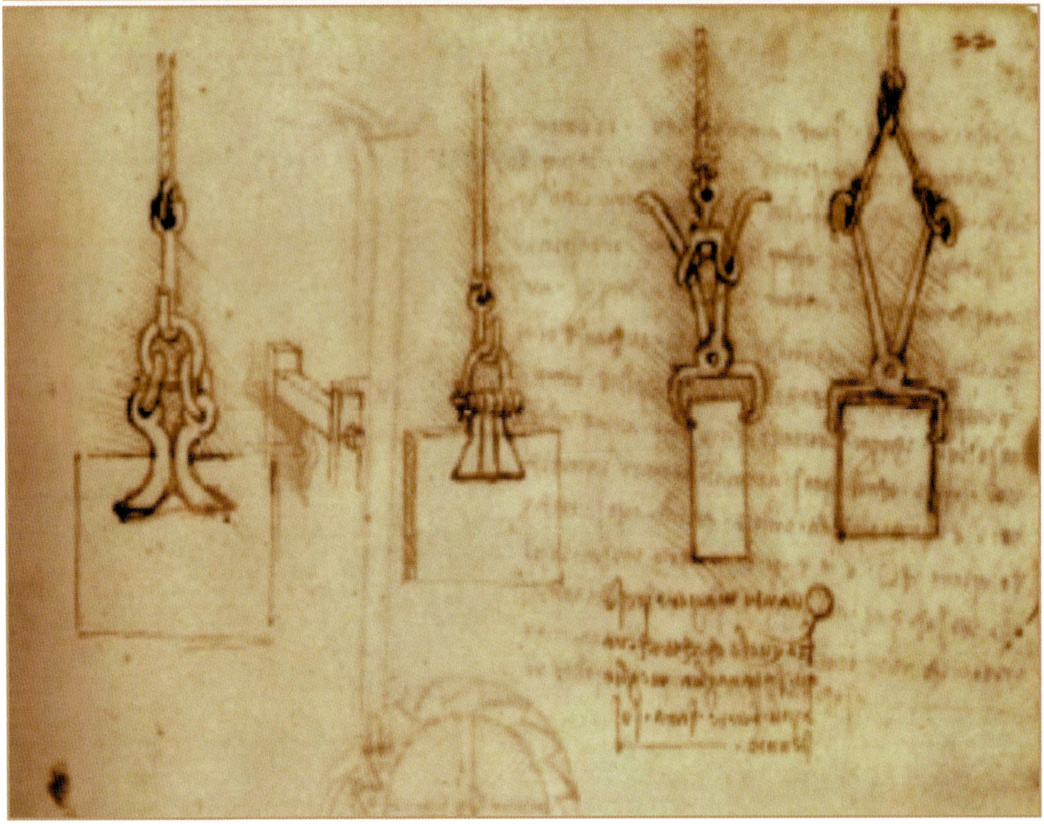

TONGS
(Pinze)
LV 36
Codex Madrid I f. 22 r

TONGS

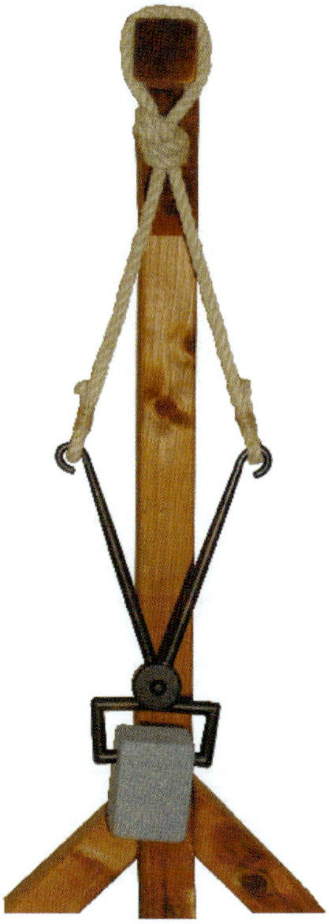

Tongs: Note how the tongs are designed to grip ever harder as the weight increases. In Leonardo's words: "The greater the weight held by this lifting tong, the better and stronger it will be supported."

VERTICAL DRILL

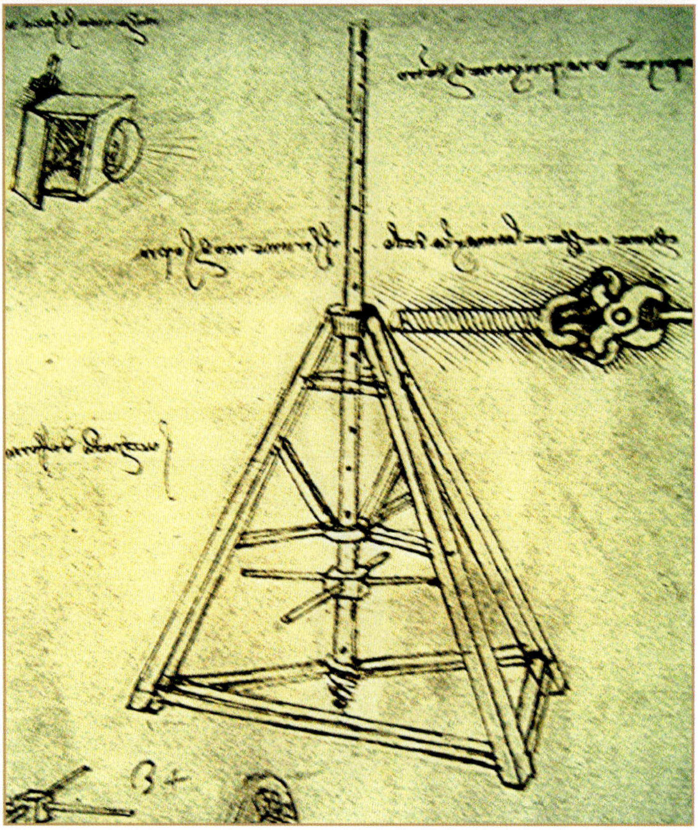

VERTICAL DRILL
(Trivella Verticale)
LV 33
Codex Atlantico f. 34 r

VERTICAL DRILL

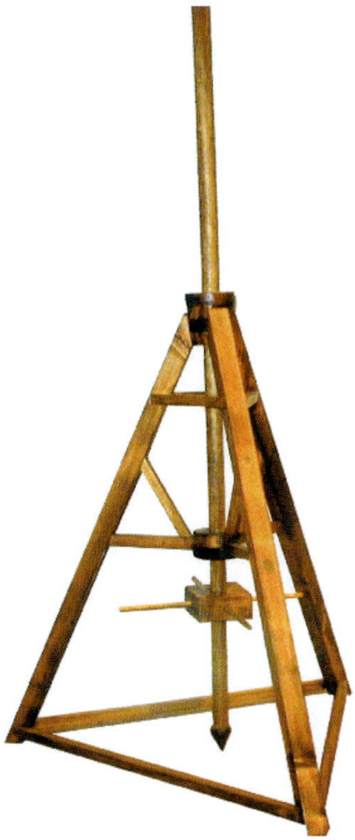

Vertical Drill: The drill bit was designed to move up and down by means of a screw activated by one or two men. The design creates a funnel-shaped space between the earth's surface and the screw head to prevent the earth from falling in on the workers.

WORM SCREW MECHANISM

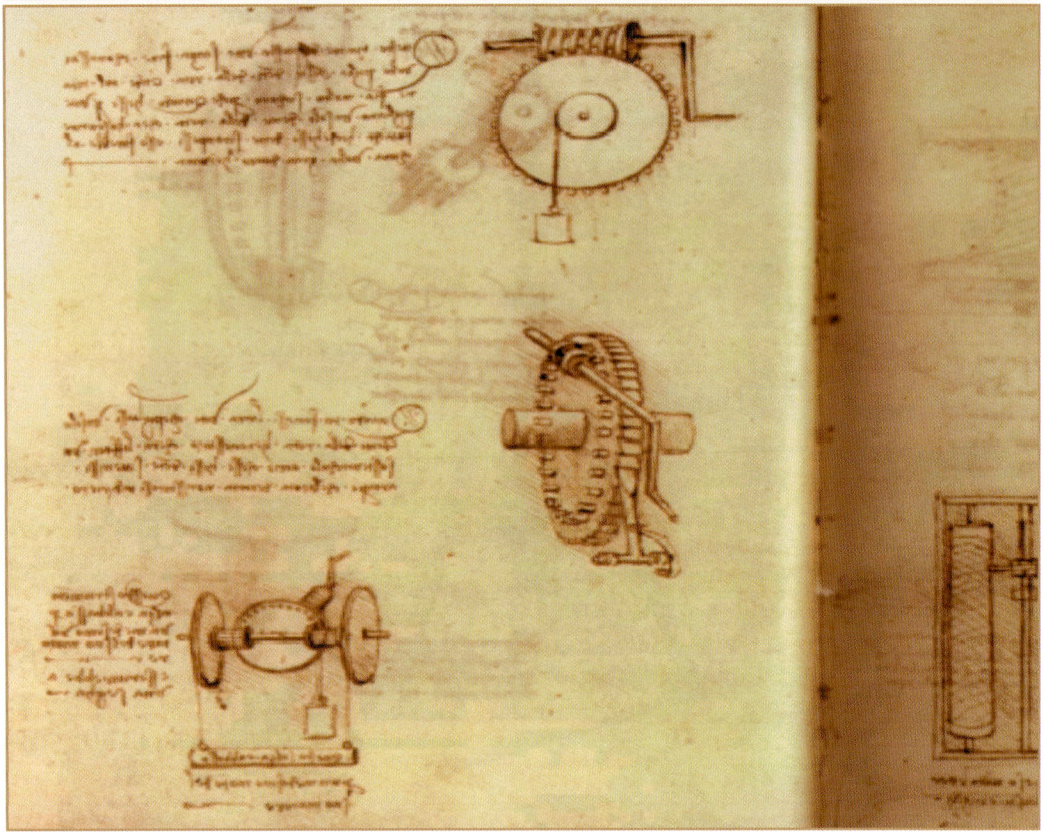

WORM SCREW MECHANISM
(Meccanismo Elicoidale)
LV 07
Codex Madrid I f. 17 v

WORM SCREW MECHANISM

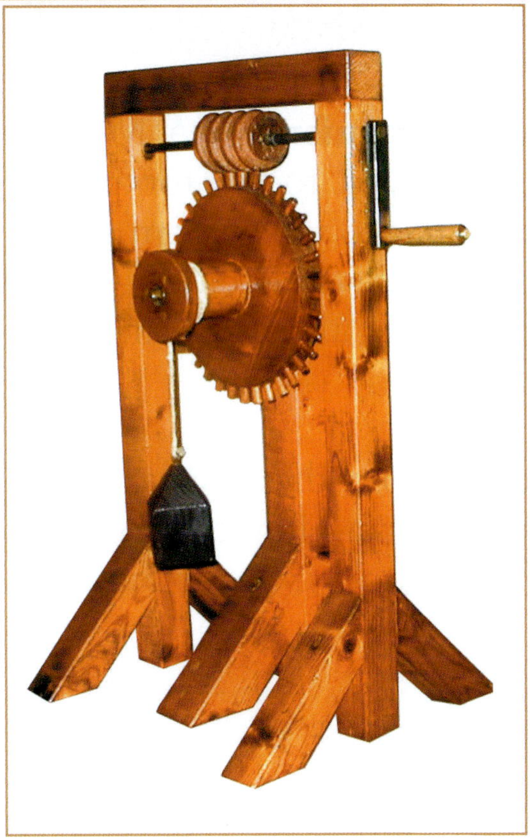

Worm screw Mechanism: Leonardo invented this "cork-screw" mechanism as a way to lift heavy objects and as a safety device: He recommended its use to prevent a wheel's inverse rotation. Because the spiral of the worm screw engages several cogs in the wheel at the same time, the force of the weight is distributed.

WORM SCREW MECHANISM — A MODERN USE

Worm screw Mechanism — A Modern Use: Worm screws are employed today for precisely the same reason that Leonardo used them at the mid-Millennium—to reduce the chance of a weight — or in the case of the tension of a guitar string — from backsliding.

As you can see from the diagram on the following page, the guitar string is wound around a post that turns a cog wheel held in place by the worm screw.

WORM SCREW MECHANISM — A MODERN USE

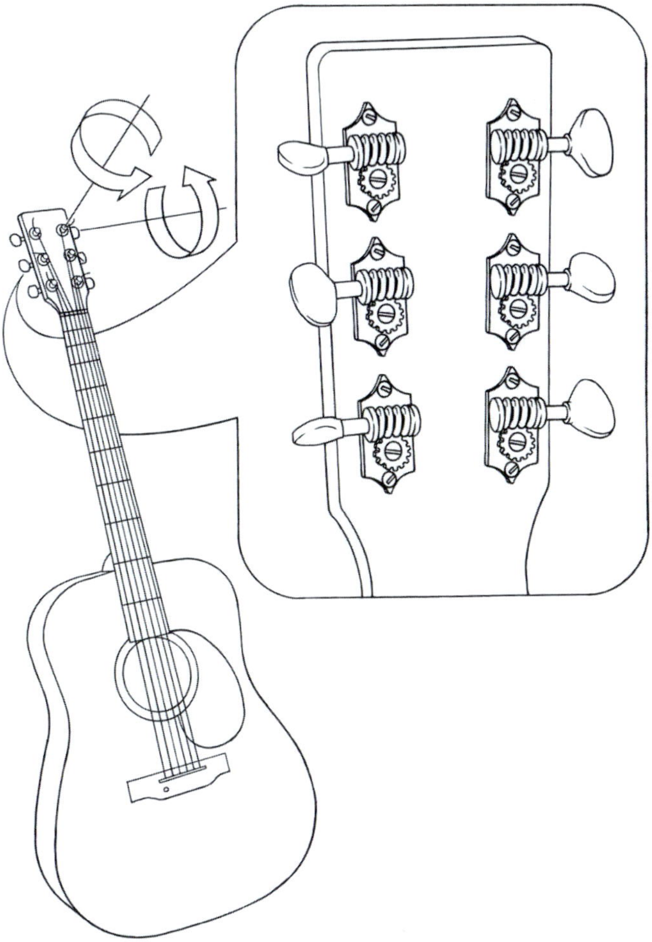

Illustration
Jamie Pfeifer Graphics

THE DA VINCI EXPERIENCE

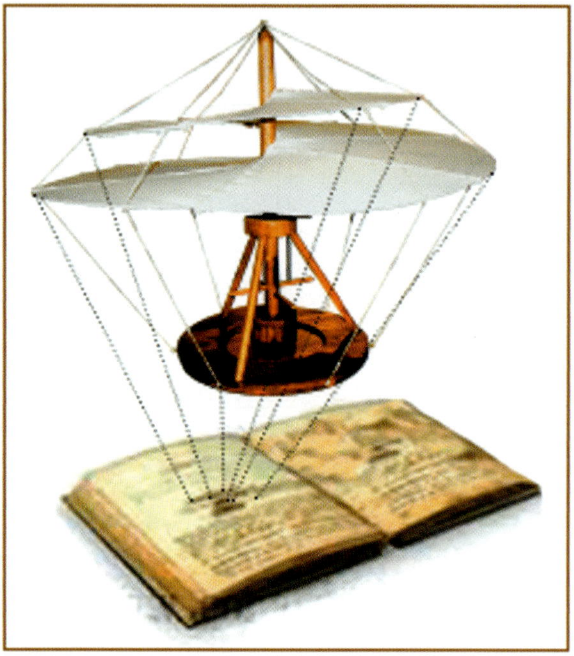

Machines Designed and Described by
LEONARDO da VINCI

OTHER ITEMS AT THE DA VINCI EXPERIENCE EXHIBIT

THE DISPLAY FLOOR PLAN

The machines described and designed by Leonardo da Vinci — as well as some of the other Da Vinci material available through the Museum of Leonardo da Vinci in Florence, Italy — are displayed in the Da Vinci Experience exhibit in three principal galleries focused around items involved in transportation, military activfities, and mechanical devices. To the extent possible, items that show Leonardo's concern with entertainment are grouped together within the mechanical gallery.

In addition, the exhibit floor consists of a small theater devoted to an orientation film about Da Vinci's life and contributions; an art gallery where replicas of some of Da Vinci's most famous artistic contributions are displayed; an educational area featuring tables to make a paper airplane, helicopter, parachute, boat and various games including a word search, crossword puzzle, maze, and a form of Sudoku called Leonardi. One wall is devoted to Leonardo's Connection — a magnetic wall display to allow kids to experiment with pipes to build an "aquaduct." A gift shop containing Da Vinci-themed books, games, and souvenir items is located near the exit of the exhibit area.

TUSCAN VIEWS

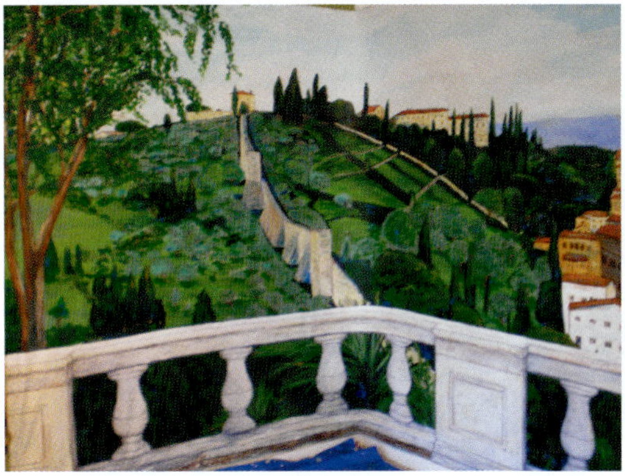

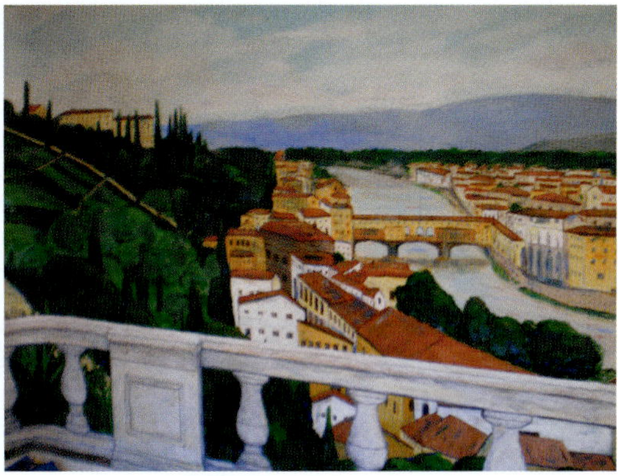

TWO MURALS PAINTED BY PATRICK HUBBARD, ORIGINALLY COMMISSIONED FOR THE AEROSPACE MUSEUM OF CALIFORNIA IN SACRAMENTO, HAVE BEEN DISPLAYED AT EACH SUBSEQUENT VENUE.

In developing the curatorial plan for the Da Vince Experience exhibit, we thought it would be appropriate to find a way to take visitors out of the familiar comforts and technological devices of the 21st century and cast them back to the 15th century in which Leonardo functioned.

We wanted to heighten appreciation of what Leonardo had accomplished, not only by virtue of his extraordinary imagination, but given the physical conditions in which he worked — without air conditioning in summer and uneven heating in winter; candlelight for many hours at night, endemic diseases, uncertain food supplies, and the like. An attempt to recreate a street in Florence did not meet our expectations. As a result, we decided to commission a muralist to paint scenes of a Tuscan city and views from the hills overlooking Florence to try to put visitors to the Da Vinci Experience exhibit in the proper mood for what they would be seeing.

THEATER

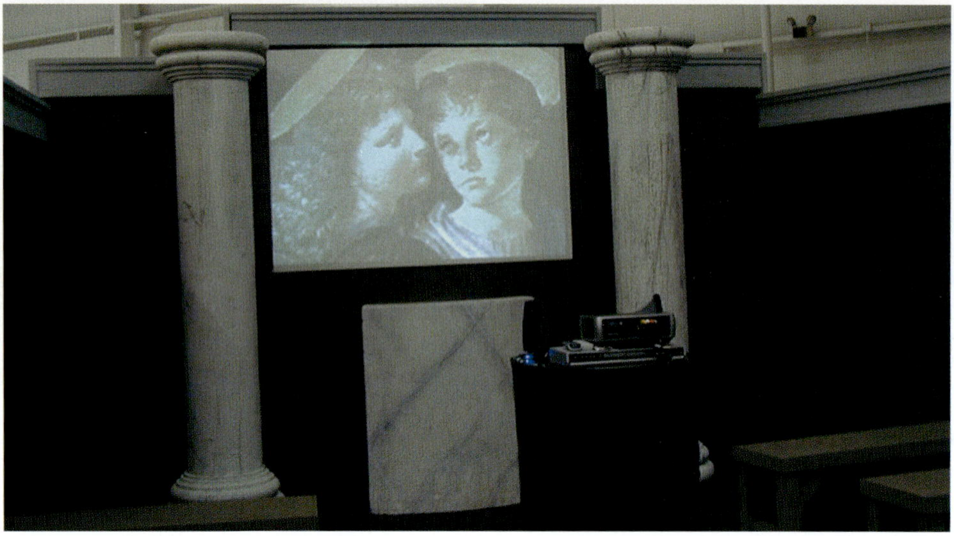

To introduce Leonardo da Vinci to visitors, an introductory film of seven minutes, dealing with his life and contributions, is shown on a continuous loop in a theater seating approximately 60 people with standing room and wheelchair space available.

The theater space can also be used for lectures, briefings, and other special events connected with the Da Vinci Experience exhibit.

The film was directed and edited by Sohrab Akhavan and written and narrated by Godfrey Harris.

LEONARDO'S ART

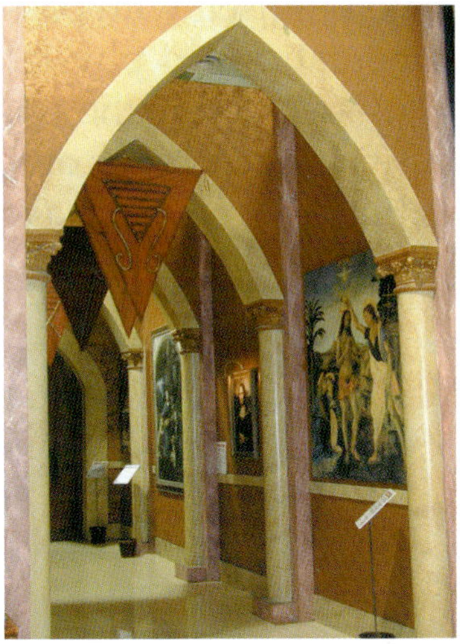

While Da Vinci's machines are a revelation to many visitors, visitors also wanted to see examples of Leonardo's artwork — the talent for which he is most renowned. How to display his art in a way that would not detract from the transportation, military, and mechanical devices that formed the core of the exhibit and yet give the reproductions their due?

During the early planning phase, we learned of a set on the Paramount Studio lot in Hollywood. It depicted a great room in a Renaissance-era Italian palazzo and had been built and used in a number of Hollywood television and film productions. The studio needed the sound stage space for other productions and wanted to sell the set. We bought it and converted it into an elegant, flag-draped hallway for the display of Leonardo's paintings. Because of its size (40′ long, 12′ feet wide and 16′ feet high), the Vestibole d'Arte could not be erected in all venues. In those cases, we have tried to capture the atmosphere of the Vestibolo by isolating the art in a space separated from the activity associated with the use of the interactive machines.

CODICES

Leonardo da Vinci dedicated himself with passion to scientific studies in anatomy, biology, mathematics and physics. The collections of his notes and sketches, known as a codex in the singular and codices in the plural, exhibit keen scientific observations and ingenious solutions to practical problems of his time as well as ideas for the future.

At Leonardo's death all his writings and sketches were inherited by Francesco Melzi, his trusted assistant and an artist and scholar in his own right. Melzi cataloged all of Da Vinci's materials, but after Melzi's death most of Da Vinci's papers were dispersed. Only an estimated one-quarter of Leonardo's known notes and sketches are believed to have survived. The codices in which they have been gathered vary in size and chronology. They are:

- Codex Arundel (1508) — in the British Museum, London.
- Codex Atlanticus (1478–1519) — the largest, with 401 sheets, held in the Ambrosiana Library, Milan.
- Codex Forster (1487-1505) — in the Victoria and Albert Museum, London.
- Codex Leicester (Hammer) (1506–1508) — owned by Bill Gates, Seattle.
- Codex Marciano (c. 1500) — in the Marciana Library, Venice.
- Codex on flight of birds (c. 1505) — in the Royal Library, Turin.
- Codex Trivultianus (1487–1490) — in the Library of Castel Sforzesco, Milan.
- Manuscripts Madrid I and Madrid II (1490–1505) — in the National Library of Madrid.
- Paris Manuscripts (labeled A to M) (1487–1515) — held in the Institute of France, Paris.
- Windsor Collection (1502-1513) — in Windsor Castle Royal Library, London

In addition, several individual drawings are held in private collections or owned by museums throughout the world.

FOSSILS

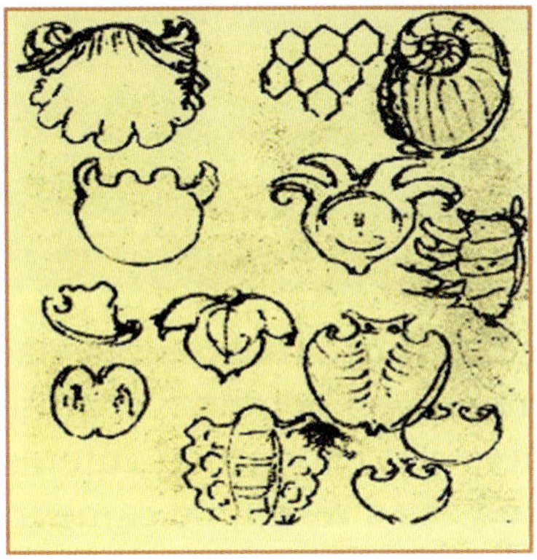

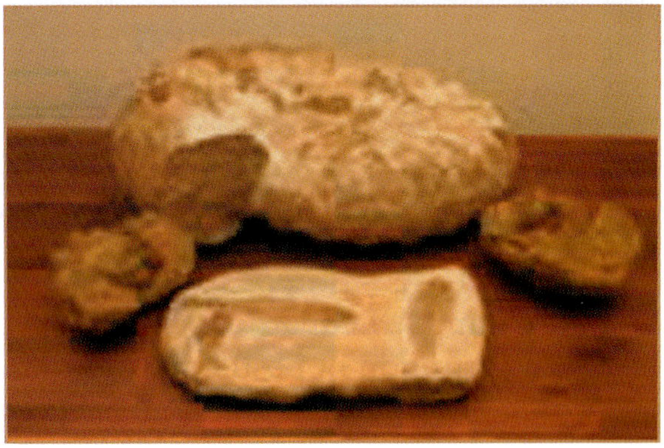

Leonardo was the first to provide what we now know to be the correct scientific explanation for the existence of fossils. His understanding of how air and water change the earth's surface over time helps to explain the geological backgrounds and panoramic views in two of his most important paintings: *Virgin of the Rocks* and *Mona Lisa*. Leonardo based his theory of geological evolution on scientific evidence provided by fossils such as shells, crustaceans, gastropods, seaweeds, and corals he found in the Alpine region north of Milan. If Leonardo were famous for nothing else, history would recognize his discovery that the Alps were once submerged under the sea and then raised by natural forces. This concept was not fully understood by the scientific community until the 1950s, when the theory of plate tectonics gained scientific acceptance. Leonardo made accurate notations on fish and leaf fossils in the Codex Leicester, Codex Atlantico, and Paris Manuscripts.

FOSSILS
(Fossili)
Manoscritto I f. 24 v 25 /Manoscritto F f. 80 r

IDEAL CITY

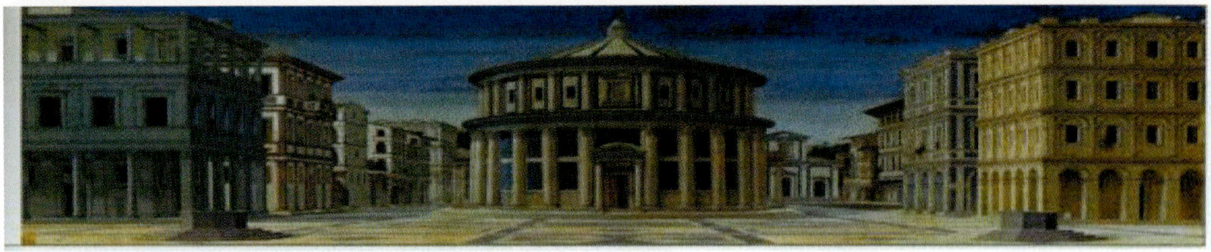

In the Middle Ages European cities had degraded so much that disease had become common. Leonardo noted that in Roman times epidemics were rare. It was the eruption of pestilence in Milan that sparked Leonardo's interest in studying how city planning might be improved. As a result, he came up with the concept of an "ideal city." He was convinced that animals' cohabitating close to humans was not hygienic. The lack of space also created many other problems. He began by studying what other Renaissance architects like Francesco di Giorgio (Siena) and Donato Bramante (Milan) had planned for their cities. He then created a new city that he said would be perfectly suited for all its inhabitants. He postulated that such a city would:

1. Be next to a river for transportation, energy, hygiene, cultivation and drinking water. The city was to be crisscrossed by navigable canals.
2. Have wide roads with their widths in proportion to the height of the buildings on either side.
3. Create a sewerage system below ground level, such as the one in Siena.
4. Build roads at ground level for heavy transport vehicles, workers, and commoners.
5. Develop elevated roads for the use of the gentry, people on horses, emergency services, and small delivery carts.
6. Require that each house have its own sewerage system and be built on aesthetic and environmental principles with access at ground level.
7. Construct warehouses for storing all necessary goods away from the houses, also with access at ground level.

IDEAL CITY
(La Citta Ideale di Leonardo da Vinci)
Manoscritto B (France)

ANNUNCIATION

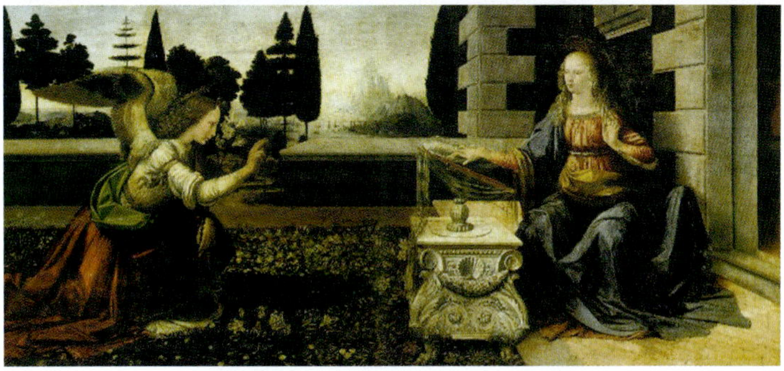

The scene in this painting is the garden of a Tuscan villa in the early 15th century. A typical architectural feature of the Florentine Renaissance can be seen in the stone work on the corners of the building. These blocks are called *ashlars* or rustication.

An angel appears to Mary to announce that She will soon become a mother. The young woman is seated in a chair wearing a red dress and covered by an ultramarine mantle and placed against the dark walls of the background masonry to allow her to be seen in rich detail. Note that the light of the picture comes from the left to illuminate the Madonna's face.

The garden, pines, cypresses, and other trees are silhouetted against the pale-blue early morning sky. In the distance, Leonardo has depicted a port city with ships and boats and even beyond that are craggy peaks leading to the vanishing point of the perspective. Scholars note that the lectern on which the Madonna rests a book and her hand was designed by Andrea del Verrocchio and obviously admired by Leonardo. All of this and other details enrich the composition, but the way that the eyes of the two figures lock on each other forces the viewer to consider the impact of the painting's central theme.

For the record, this painting is held by the Uffizi Gallery in Florence.

BAPTISM OF CHRIST

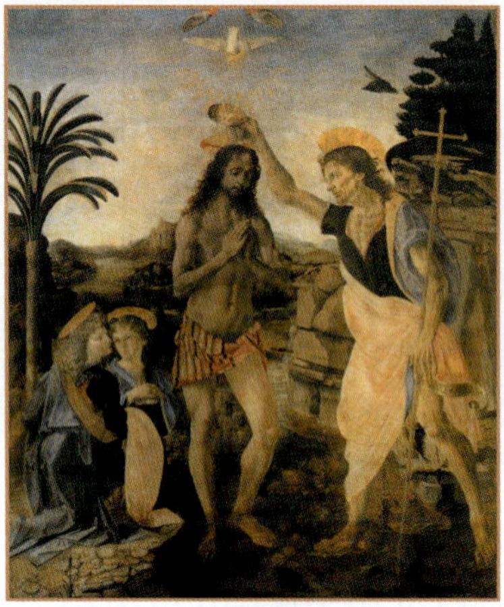

Andrea del Verracchio was given the commission to paint this picture in the 1470s. As was the custom of the time, he invited apprentices in his studio to participate in the work. Del Verracchio undoubtedly laid out the arrangement of the figures and settled on the size and material to be used—oil and tempera on a panel measuring 151 by 157 cm (59.25 x 61.75 inches). The painting may be seen today in the Uffizi Gallery in Florence.

Leonardo was given the principal honor of painting one of the two angels in the bottom left corner. Study them closely, then see if you can determine which of the two was most likely to have been painted by him.*

In addition, scholars have determined that Leonardo's unique technique is visible in the figure of Christ, the background landscape and trees, and the water in which the feet of the two principal subjects are submerged.

When del Verracchio saw Leonardo's work on this painting, he was said to have been so impressed with what his teenage student had produced that he is reported to have proclaimed that he would never paint anything ever again. As far as we know, he didn't.

---

* Everyone agrees with author Maria Teresa Zanobini Leoni that "the angel on the left was certainly by [Leonardo's] hand."

GINEVRA DE'BENCI

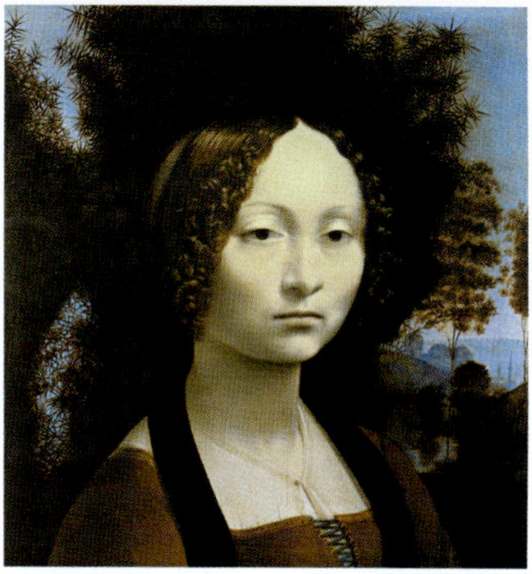

Ginevra de'Benci is one of Da Vinci's early works, painted around 1474 when Leonardo was still at Andrea del Verrocchio's bottega. Historians generally consider the portrait to have been commissioned to celebrate her marriage to Luigi Niccolini, someone twice her age. She was a renowned intellectual, seventeen at the time the painting was completed. On the reverse of the wood panel on which the portrait is painted is a heraldic motif consisting of a sprig of juniper encircled by a wreath of laurel and palm, with the motto: "Beauty adorns Virtue."

As with all of Leonardo's portraits, there are disagreements over this one and some researchers feel it may have been commissioned by Bernardo Bembo, the Venetian Ambassador to Florence. Bembo and Ginevra, both married, were known to have had a platonic affair, quite an accepted convention of the day. One of the most intriguing elements of this painting is the fact that a juniper tree is also seen directly behind the subject. The Italian word for juniper is "ginepro." The inclusion of this plant may be a little inside symbolism from Leonardo to mark his subject's first name and to honor a Renaissance symbol for chastity.

The painting today is the only Leonardo painting in North America. It is held by the National Gallery of Art in Washington, D.C. The painting is also noteworthy because Ginevra's arms and hands were lost when a bottom panel was removed at some point in the past.

LADY WITH AN ERMINE

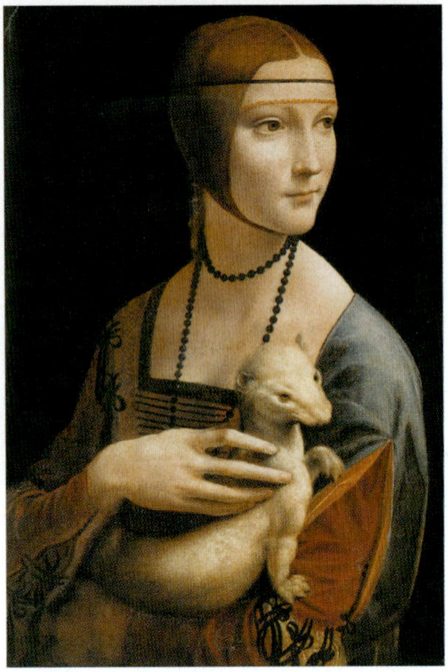

The *Lady with an Ermine* was painted by Leonardo between 1488 and 1490. Like many other Leonardo portraits, this was executed in oil paints on a wood panel. The painting is also relatively small: 39 by 54 centimeters or about 15 by 21 inches.

No one knows the name of the woman who sat for this portrait, although most scholars think it was Cecilia Gallerani, a favorite of Duke Ludovico Sforza. The ermine she holds is thought to be a symbol of chastity.

This painting is one of the examples of the 3-D effect Leonardo was able to achieve — the dark background and the lighting on the subject gives a great feeling of depth and volume. Leonardo's remarkable skill in portraying hands is particularly interesting in this portrait, with the elongated fingers seeming to mirror the long body of the ermine. But note the lady's left hand. Is it unfinished as with other da Vinci projects, a subtle comment on attitudes toward lefthandedness, or just crudely done to emphasize the elegance of the right hand?

THE LAST SUPPER

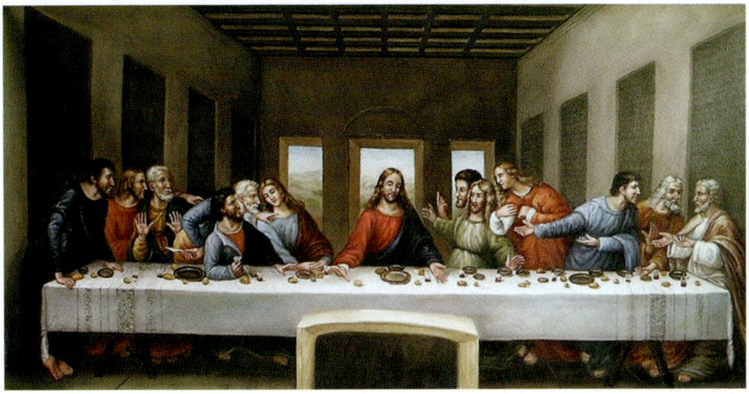

During the period of 1496 to 1498 Leonardo painted the *Last Supper* on the wall of the Refectory of the Dominican Convent of Santa Maria delle Grazie in Milan. In the painting, Leonardo portrays the moment when Jesus said: "One of you shall betray me." The different gestures and expressions of the disciples at that instant are meant to show that the apostles realize there is a traitor among them. To Christ's left is St. John, who is taken with grief and questioned by Peter. Judas Iscariot is pictured calm and innocent, while he holds a bag containing thirty pieces of silver.

The painting was originally executed on a specially prepared stucco base to allow Leonardo to change aspects of the painting as it progressed. Unfortunately, the fresco began to deteriorate a few years after the painting was completed. By 1556, it was half ruined. In 1652, the monks cut away a part of the fresco, including the feet of Christ, to make a doorway. In 1726, a Milanese painter named Michelangelo Belotti was given the first job of restoring it. It is believed that Belotti used an astringent that temporarily revived the colors but later left them even duller than before. It was not until 1908, after several other attempts to restore the piece, that one Professor Luigi Cavenaghi completed a restoration that has preserved what remained of the mural.

MUSICAL NOTES HIDDEN IN THE LAST SUPPER

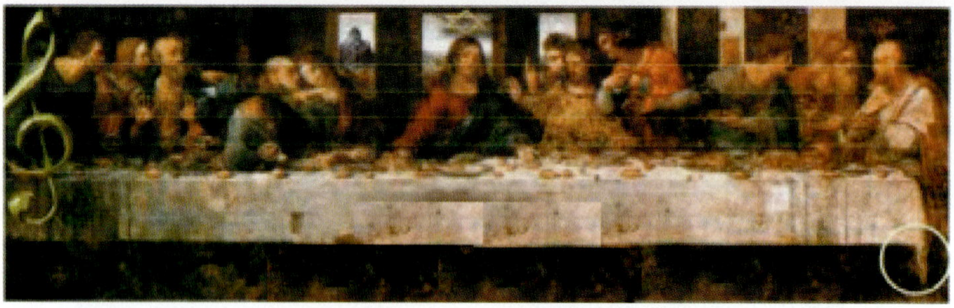

When Italian musicologist Giovanni Maria Pala drew the five lines of a musical staff across *The Last Supper*, he noticed that the loaves of bread, fruit, and hands of Jesus and the Apostles could each represent a musical note. If these musical symbols are played on an instrument in the position they appear, they produce the sound and rhythms of a requiem. Pala's research is detailed in a book entitled, *Leonardo da Vinci's Musical Gifts and Jewish Connections.*

While Pala's thesis may be the result of sheer happenstance, we do know that Leonardo liked to compose riddles and hide symbols within his work. Renaissance intellectuals also believed "a proficiency in music went with the study of art and architecture." It is not surprising, then, that Leonardo's first royal appointment was as a musician at the Milanese court. He played lyre (or more properly, a *viola da braccio* — a six-string prototype of the modern violin) and was said to sing beautifully. He is also known to have designed and probably built a number of musical instruments during his career, yet no musical composition has been found and only poems that might have been set to music have been discovered in some of his notes and sketches.

Pala believes that the knot tied into the table cloth in the lower right corner of the fresco is a representation of the knots at the fringes of Jewish prayer shawls — Leonardo's subtle reminder to his Dominican hosts and the principal advocates of the Inquisition (the campaign to discover and suppress heresy) that all 13 men depicted in *The Last Supper* were, in fact, Jewish.

LA BELLE FERRONNIÈRE

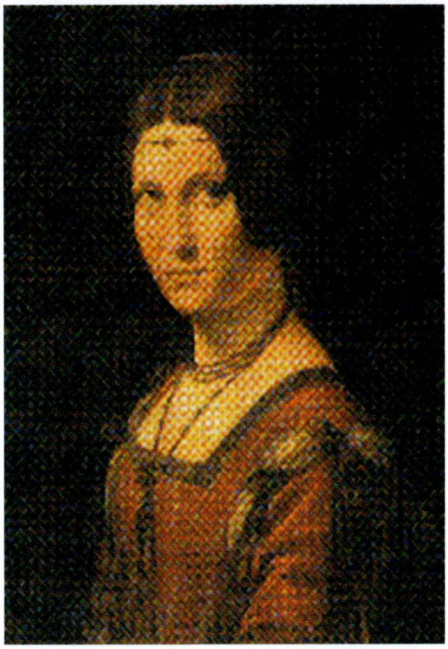

Scholars use this small oil on panel (44 x 62 cm or 17 1/4 x 24 1/2 inches), now in the Louvre, to illustrate Leonardo's geometric approach to portraiture. Here the head is an ellipsoid and the eyes are almond cut shapes. The hair emphasizes the curves of the face to create an oval. The ribbon hemming the bodice sets the size of the lady's shoulders.

Note the pose. The body is three-quarters to the viewer, the face full, but the eyes are averted to the side. It has the appearance of what might be found in a sculpture and the plain background suggests that that is what Leonardo might have had in mind. The pose and background of the portrait of *St. John* bears similarities to this.

About the name. The title of the painting, *La Belle Ferronnière,* refers to the wife of an ironmonger. She is most probably the mistress who succeeded the *Lady with an Ermine* in the affections of Duke Ludivico Sflorza. She is believed to have nothing to do with an ironmonger. The confusion arises because the Polish owner of *Ermine* erroneously subtitled *her* "la belle ferronnière" when it was assumed *she* had been the mistress of the King of France.

MONA LISA

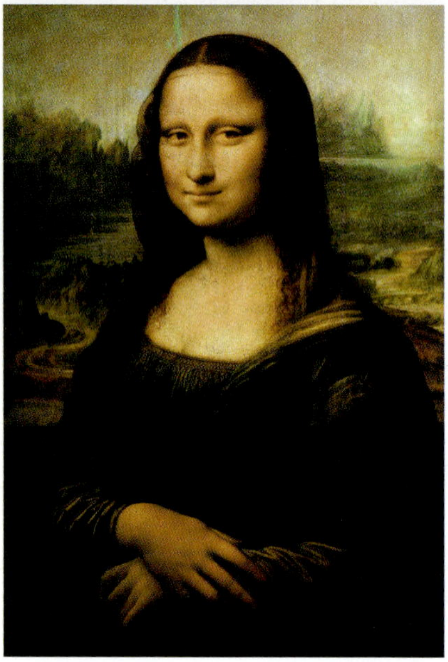

Early in 1503 Leonardo returned to Florence and began work on the *Mona Lisa* ("Madam Lisa" in English), another small (44 x 62 cm or 17 1/4 x 24 1/2 inches) oil on (poplar) wood painting. It is generally thought to be a portrait of Lisa Gherardini, the wife of Francesco del Giocondo, a wealthy merchant. Leonardo worked on the portrait on and off for the rest of his life. The study of the *Mona Lisa* has occupied the attention of scholars for years. In the florid words of one 19th century anonymous writer:

> *Whoever shall desire to see how far art can imitate nature may do so to perfection in this [portrait], wherein every peculiarity that could be depicted by the utmost subtlety of the pencil has been faithfully reproduced.*

The half-length portrait is perhaps most famous for how the woman's gaze meets the viewer's and her enigmatic expression. Recent speculation about the meaning of the expression, based on a laser analysis of how the painting was created, suggests she has just learned of her pregnancy and is wearing a special garment to signify her condition.

THE NUDE MONA LISA

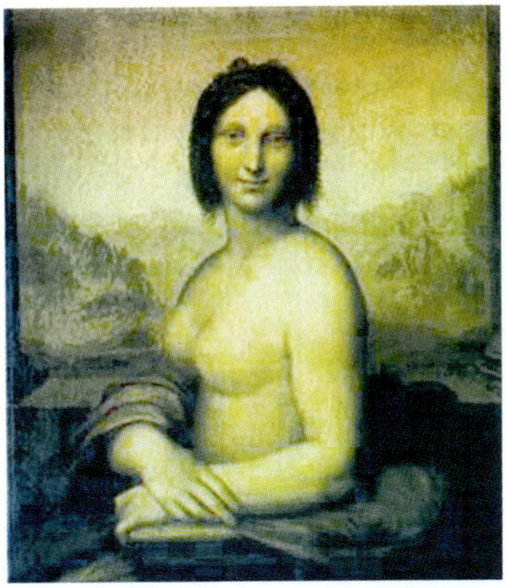

This painting is on display at the Museo Ideale in Vinci, Italy, after being hidden for years in a private library. While resembling in many ways important aspects of Leonardo's renowned painting of the wife of wealthy Florentine merchant, Francesco del Giocondo, it is probable that this particular painting was done by a student of Leonardo from an idea the master himself had conceived. Some see a similarity in the face and hair of this *Monna Vanna* to Leonardo's *Leda and the Swan*, a painting of a mythological event that survives in his sketches and in treatments by others.

The nude Mona Lisa was acquired by Napoleon Bonaparte's uncle, Cardinal Joseph Fresch, from a rich aristocratic Roman family. Cardinal Fresch was Napoleon's Ambassador to the Vatican and an art collector of considerable fame. In documents exchanged at the time of sale, the subject of the painting is said to be the mistress of Francis I of France, and the artist is identified as Leonardo da Vinci. After the Cardinal's death in 1839, Napoleon III was thought to have owned the painting prior to it going to the collection of a descendant of Napoleon's brother.

Scientific and artistic investigations are concurrently underway to date this work and to try to identify the actual painter.

PORTRAIT OF A MUSICIAN

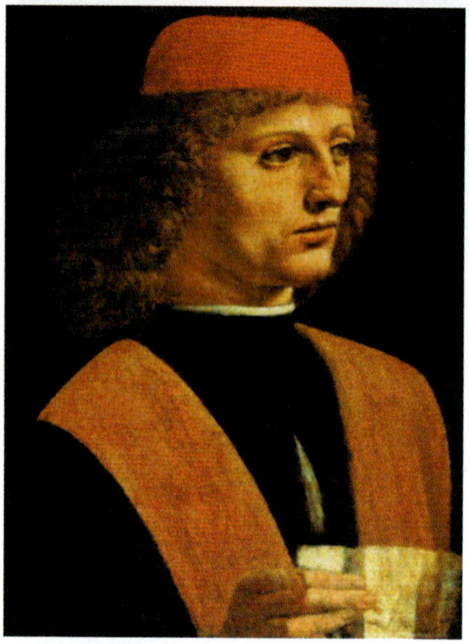

This painting shows a lot of Leonardo's wonderous talent as a painter. The dark background pushes the subject toward the viewer, giving the painting a three-dimentional presence. The orange bands of the cloak, the white collar underneath the tunic, and the soid red cap add volume and definition to aspects of the painting that add depth to the subject.

Note the musical score with visible notes in the bottom right and the delicate hands of the musician. Although we would think of a musician as primarily someone who plays an instrument, Leonardo's *Musician* was most probably a singer, given his full neck and powerful chest.

The painting was executed between 1482 and 1484 and like most of his other small works is done in oil on a wood panel, (31 x 43 cm or 12 1/4 x 17 inches). The original of this work can be seen in the Ambrosian Gallery in Milan.

BELLA PRINCIPESSA

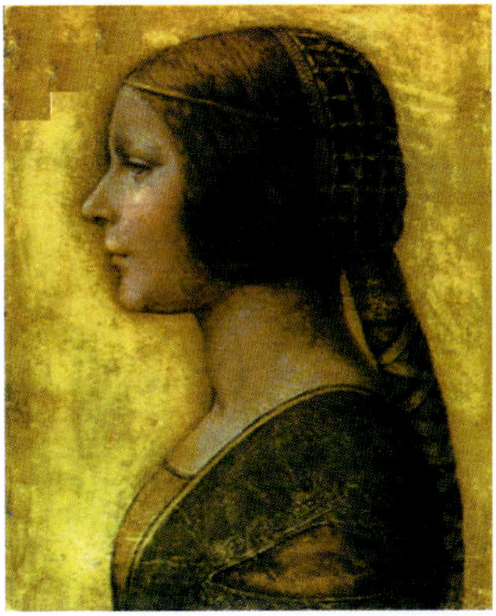

This portrait may have been executed by Leonardo da Vinci, perhaps at an early point in his career. In October 2009, a Canadian forensic art expert announced his discovery of both a palm print and a fingerprint, known to be Leonardo's, in the upper lefthand corner of the chalk, pen, and ink drawing. The small (9 x 13 inches or 24 x33 cm) rendering of a young woman was done on vellum and then mounted on wood.

The portrait had been catalogued by Christie's New York as "German, early 19th century" and was sold for $19,000 in 2007 to a Swiss collector. Because it could be the first major Leonardo find in the last 100 years, some have suggested that the drawing might have a worth well more than 1000 times the sum paid.

While other experts try to determine the *Bella Principessa's* provenance, form your own opinion as to whether this portrait has the characteristics of a true Leonardo. How many other paintings did he do in which his subject was in *full* profile? When else did he work on vellum and then mount the finished work on wood? Could Leonardo's finger and palm prints merely indicate that he had handled the painting, at some point, rather than establish definitively that he had painted it himself? Experts are continuing their investigation to try to find out.

ST. JEROME

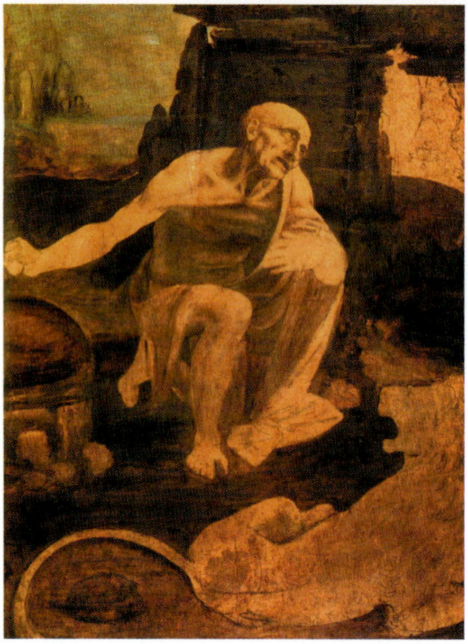

Saint Jerome is credited with having translated the Bible into Latin. As a result, he became the patron saint of translators as well as a symbol of a man dedicated to the study of God. He is often pictured, as here, with a friendly lion nearby, having previously removed a thorn from the lion's paw. In this 1482 work, Leonardo produces a landscape that is similar to the background seen later in the Mona Lisa. The pale light filtering into the monastic cave from the right was imaginatively achieved by simply leaving the gesso (the layer used to cover the raw wooden panel) unpainted. St. Jerome's anguished expression is thought to reflect the pain he imagined Jesus experienced, having been stimulated by the Crucifix seen to his left. Note also the tilt of his head, the stiff positioning of his right hand, and the thrust of his knee. These are conveying a sense of movement — a technique copied by generations of future artists.

The painting — 108 x 73 cm or 42.5 x 29 inches in the Vatican Picture Gallery — has an unfinished quality, the result of Leonardo's habit of initially sketching subjects in paint (rather than in charcoal or lead) and using drafting techniques to illuminate a subject. Leaving work unfinished, of course, would be one of the more intriguing aspects of Leonardo's life story.

ST. JOHN THE BAPTIST

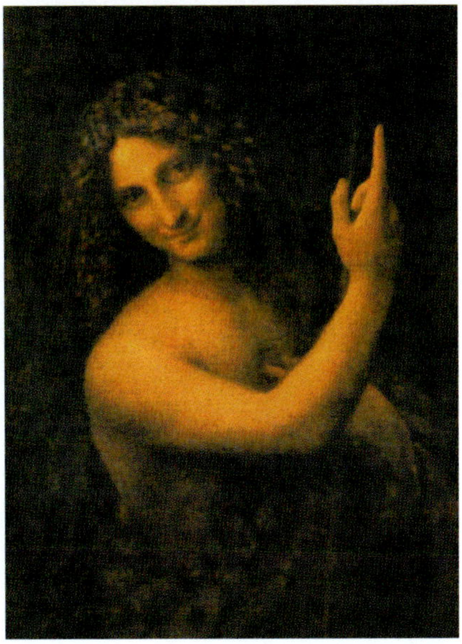

This small portrait of St. John the Baptist was done late in Leonardo's life while the guest of King Francis I of France. It may well be his last work because he suffered a stroke in the same year that this painting was completed. It is an oil painting on a panel and measures only 57 x 69 cm or 22 1/2 x 27 inches. Like every other painting that Leonardo had in his possession at the time of his death in 1519, this one remained with the King of France and now hangs in the Louvre in Paris.

Leonardo is known to have made done several St. Johns and it may well be that because of Leonardo's stroke, this was completed by his assistants, Francisco Melzi and/or Giacomo Caprotti ("Salai.") In the case of this St. John, it would seem that Leonardo created the pose. Note that the model is turned three-fourths to the front, but the head is facing fully forward — a juxtiposition used in *La Belle Ferronnière*. The gesture of the finger pointing to heaven suggests a belief in the coming of Christ. Given the shadowing of the face, scholars assume it was done by Leonaardo's hand. Some think that it may also be a self-portrait of Leonardo as he remembered himself at a younger age, given the similarities between this face and the boy who became Andrea del Verrocchio's *David.*

VIRGIN AND CHILD WITH ST. ANNE

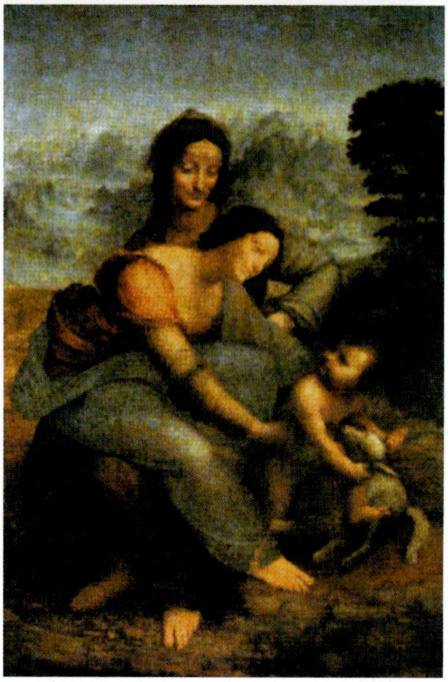

This painting, also referred to by some scholars as *St. Anne, the Madonna, and the Christ Child with the Lamb,* is riveting, not only because of the arrangement of the people in the scene, but because of the fanciful background. This relatively large oil (112 x 168 cm or 44 x 66 inches) is in the Louvre and was completed in 1510. The movement implied in the painting — as St. Anne reaches for the infant, as the lamb struggles uncertainly — adds wonderful drama to the composition.

Leonardo apparently worked on the arrangement in several detailed attempts to get it right. A charcoal and white lead cartoon, dated 1500-1501, is in the National Gallery in London. That cartoon included a representation of St. John the Baptist, arranged in a square. Leonardo apparently didn't like that and tested at least two other arrangements. Eventually he decided to compose the subjects in a triangular or pyramidic configuration with a lamb substituting for St. John. In late 2008, curators discovered a "a water stain" on the back that turned out to be three faint drawings of a horse, a skull, and a child and lamb that might have been a test of what Leonardo wanted for the other side.

VIRGIN OF THE ROCKS

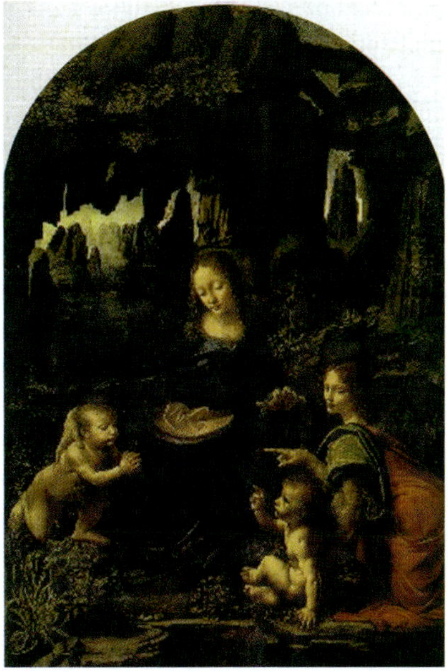

In 1483, Leonardo signed a contract to paint the *Virgin of the Rocks* together with the Da Pedris brothers. A letter in the State Archives of Milan from Giovanni Ambrogio da Pedris and Leonardo da Vinci to the Duke of Milan asks the Duke to intervene in a dispute with the Brotherhood of the Conception. The disagreement was over the value of works of art furnished for the chapel of the Brotherhood, including the *Virgin of the Rocks*.

Dispute aside, the painting is renowned for the "compositional matrix" that puts the Virgin (at the center) with the Christ Child (on the left) and the infant St. John and the angel (on the right). As Maria Teresa Zanobini Leoni writes in her book, *Leonardo da Vinci,* the figures are linked by a "complex interplay of direct and reflected light" while the "rendering of the … rock formations … [reveals Leonardo's understanding] of nature."

VITRUVIAN MAN

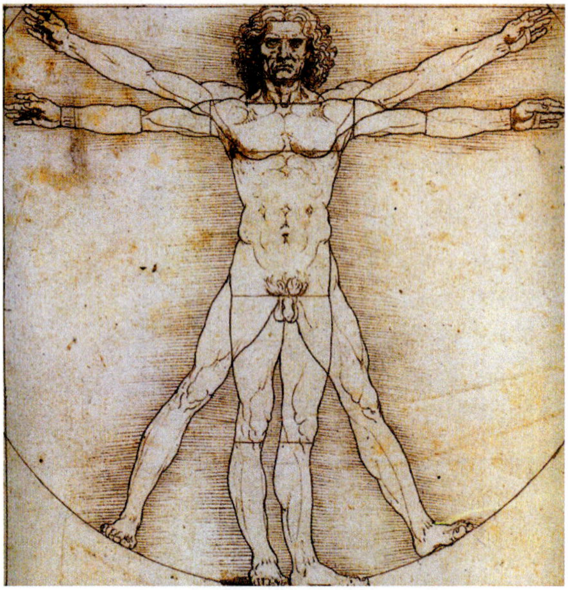

Leonardo believed nature created perfect machines. The human body was one of those machines. His famous drawing known as the *Vitruvian Man* (named for the Roman architect's ideas of golden proportions) shows the relationship between a person's outstretched arms and his height — a blending of Leonardo's interest in both art and science. Because Leonardo was among the first to use mathematics to understand human anatomy,other relationships emerged:

- A palm is the width of four fingers.
- A foot is the width of four palms.
- A man's height is equal to 24 palms.
- The distance between the top of a man's head and the base of his chin — or the space between an elbow and an armpit — is equal to one-eighth of a man's height.
- The maximum width of a man's shoulders is equal to one-quarter of his height.

The symbolism of the Vitruvian Man is so universal that Italy uses a version of Leonardo's design as the symbol on its 1 euro coin.

BATTLE OF ANGHIARI

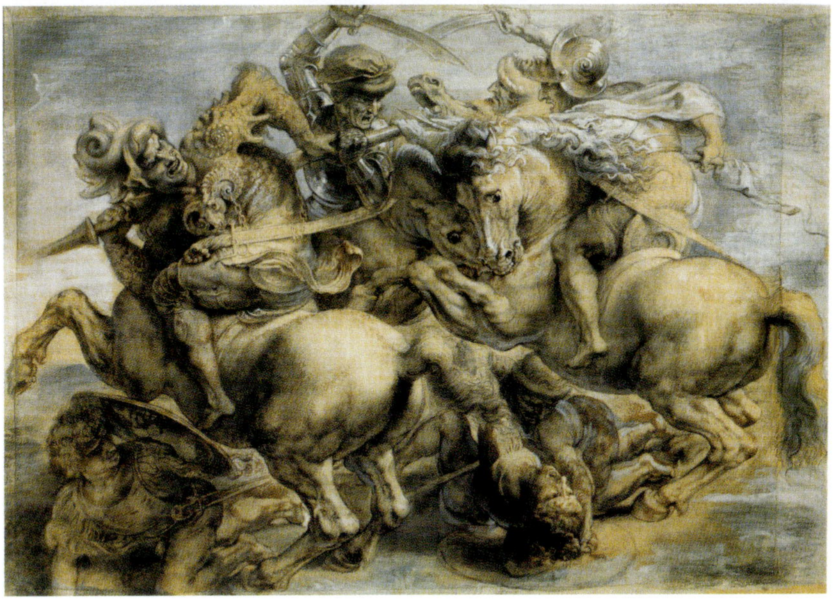

The Italian Ministry of Culture has approved a continuation of the search for the *Battle of Anghiari,* a fresco Leonardo da Vinci started in 1503 to depict a victory over the Milanese. Some months later Michelangelo began the *Battle of Cascina* on the opposite wall of the Great Hall in the Palazzo Vecchio in Florence. Both pieces were commissioned to commemorate the 10th anniversary of the founding of the Republic of Florence. Neither finished his painting, but the central element of Leonardo's mural — called *Fight for the Flag* — was much admired by contemporary artists who saw it; the work depicted here in chalk, pen and ink is a 45.2 x 64.7cm (about 18 by 25 inches) 1603 copy of an engraving by Peter Paul Ruben owned by the Louvre in Paris.

During a major renovation of the Palazzo by Georgio Vasari, the famed art historian, the unfinished Da Vinci work was thought to have been covered by a wall, but preserved behind. A search for the cocoon containing the painting was begun by Florentine preservationists some 30 years ago, but suspended. The investigation of the Palazzo's walls by non-invasive techniques is continuing and the art world will wait with great anticipation should any portion of this renowned work be found and perhaps put on display.

LEONARDO SELF-PORTRAIT

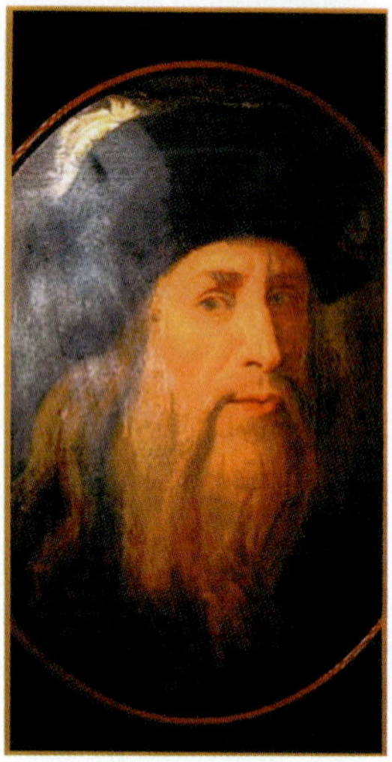

This is a retouched image of a damaged wood panel portrait of Leonardo da Vinci uncovered in early 2009 in a library in southeastern Italy. Leonardo was known to have traveled to this area during his lifetime where an aristocratic Florentine family had a home.

The 17 x 23.6 inch (60 x 44 cm) portrait is now being intensively studied and scientifically reviewed as a possible self-portrait of Leonardo in the prime of his life — perhaps in his late 30s or early 40s. The painting is said to show no preparatory work — typical of Leonardo's approach — but several later additions after the main work had been completed — the type of fussing that Leonardo was also known to do.

A handwritten inscription on the back, in reverse script, reads *pinxit mea* — indicating either an authentic self-portrait, a prank, or the cunning note of a forger trying to enhance his fake. Again Leonardo presents an intriguing mystery to be solved nearly 500 years after his death.

APPENDIX

ACKNOWLEDGMENTS

The following individuals deserve our thanks and sincere appreciation for their contributions to the Da Vinci Experience. Without their able assistance, Kenny Katz and Godfrey Harris would have not been able to succeed:

FOR DA VINCI EXPERIENCE, LLC

Sohrab Akhavan
William P. Butler
Christopher Carroll
Gregory Cherninsky
Robert Crawford
Arthur E. Detman
Barbara DeKovner-Mayer Harris
Kennith L. Harris
Mark B. Harris
Robert Jamieson
Shailesh Joshi
Katy Katz
Vanessa Katz
Josh Landon
Michelle LeFort
Harlan Levinson
Thomas Mankowski
Michael Mathews
Michael Meltzer
Marcia McCready
Penelope Morensen
Gabriele Niccolai
David O'Brien
John Perry
Carmen Santana
Ursula Santos
Joe Scarna
Richard Sherer
Adolf P. Shvedchikov
Phil Tendler
Julie Trejos
Aaron Vida

ABOUT TECKNOART

The Da Vinci Experience exhibit in the United States would not have been possible without the trust, contributions, and dedication of Gabriele Niccolai and Luigi Rizzo. Both of these men are Italian by birth, the former in Tuscany and the latter in Venice. Gabriele supervises the Museum of Leonardo da Vinci in Florence; Luigi manages Tecknoart Australasia, the company that provides a traveling version of the Florence display.

Only about one-quarter of the notes and sketches believed to have been made by Leonardo da Vinci have been found. When a group of them were seen by scholars in the late 1950s they discovered a number of Da Vinci's concepts for mechanical devices. Museums wondered if they could be brought to life. A young craftsman in Florence—someone who knew the Tuscan dialect, had the patience to read it backwards and could understanding Leonardo's complex notes and diagrams—was asked to create replicas using the same materials and techniques that would have existed in Leonardo's time. Carlo Niccolai was eventually joined in this enterprise by his son, Gabriele, who brought computer technology to the task of recreating even more of Leonardo's machines. They founded the Museum of Leonardo da Vinci in Florence which has recently been recognized by eminent Da Vinci scholar Professor Carlo Pedretti and the Da Vinci Foundation as the authentic museum of Leonardo da Vinci.

Luigi Rizzo, a citizen of Australia, met Gabriele Niccolai while living and working in Florence. The Niccolai family had previously made arrangements to send replicas to Vienna, Dubai, Venice and other cities. Rizzo offered to expand the Niccolai family's contribution to understanding Leonardo da Vinci to the Pacific Rim. Gabriele accepted the offer and together they created a large traveling exhibit for multiple venues in New Zealand and Australia. When Rizzo asked his American colleague, Godfrey Harris, if he could find suitable sites for these same Da Vinci replicas in the United States, he agreed to try. Harris and his long-time associate, Kenneth Katz, soon made arrangements to have the Da Vinci Experience exhibit in Palm Springs and Sacramento, California and Henderson, Nevada. The exhibit is expected to travel to Orem, Utah, and San Diego, California, in the future. The U.S. exhibit has expanded on its New Zealand and Australian origins by providing high quality reproductions of Da Vinci's most representative paintings in a unique setting. The Da Vinci Experience has also contributed an engaging children's educational and craft center as part of the display.

No matter the venue, visitors comment on the way the replicas and the art are presented in such a way to expand their appreciation of Leonardo da Vinci's contributions.

BIBLIOGRAPHY

The following books were useful in conceptualizing elements of the exhibit and in preparing this catalog:

Brown, Dan, *The Da Vinci Code*, Doubleday, 2003

Burns, Edward McNall, *Western Civilizations*, W.W. Norton, 1955

Dann, Jack, "Da Vinci Rising," *Nebula Awards 32*, Science Fiction Writers of America, 1998

Cianchi, Marco, *Leonardo The Anatomy*, Giunti, 1998

Dickens, Emma, editor, *The Da Vinci Notebooks*, Arcade Publications, 2005

Freud, Sigmund, *Leonardo da Vinci and a Memory of His Childhood,* Standard Edition, 1916.

Hobson, Charles, *Leonardo Knows Baseball,* Chronicle Books, 1991

Konigsburg, E.L., *The Second Mrs. Gioconda,* Simon & Schuster, 2003

Leoni, Maria Teresa Zanobini, *Leonardo da Vinci*, McRae Books, 2002

McCurdy, Edward, *The Mind of Leonardo da Vinci*, Dover Publications, 2005

Nicholl, Charles, *Leonardo da Vinci*, Viking, 2004

Pala, Giovanni Maria with Loredana Mazzarella, *Leonardo da Vinci's Musical Gifts and Jewish Connections,* The Americas Group, 2010

Shalev, Baruch, *100 Years of Nobel Prizes*-Fourth Edition, The Americas Group, 2005

Starnazzi, Carlo, *Leonardo Codexs & Machines,* Cartei & Bianchi, 2005

Suh, Anna H., *Leonardo's Notebooks*, Black Dog & Leventhal, 2005

Teisch, Jessica with Barr, Tracy, *Da Vinci for Dummies*, Wiley Publishing, 2005

Wolk-Simon, Linda, *Raphael at the Metropolitan,* Metropolitan Museum of Art, 2006

Zöllner, Frank, *Leonardo da Vinci,* Taschen, 2006

INDEX

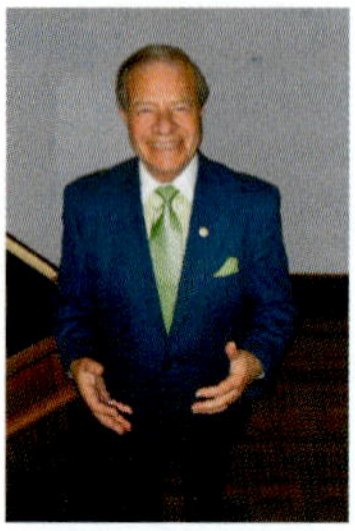

GODFREY HARRIS has been a public policy consultant based in Los Angeles, California, since 1968. He began consulting after serving as a university lecturer, U.S. Army intelligence officer, U.S. foreign service officer with the Department of State, an organizational specialist in President Lyndon Johnson's Executive Office, and as a program manager for an international financial company in Geneva. As President of Harris/Ragan Management Group, Harris has focused the firm's activities on projects that offer alternative solutions to matters of community concern.

Given the number and variety of projects he has been involved with over his 40-year career, it was not unusual for him to accept the challenge of curating an international exhibit of the machines designed and described by Leonardo da Vinci. This has involved not only finding suitable venues, but envisaging how the exhibit should be organized, preparing the signage to be used, scripting the introductory film, creating and finding items for sale in the Gift Store, specifying the objectives of the exhibit in terms of their entertainment and educational goals, as well as a selection of Da Vinci's art to complete the experience. In addition, Harris has been involved in the commissioning, editing, indexing, and publication of two new books about Leonardo da Vinci.

This fifth edition of the Da Vinci Experience Exhibt Catalog is the 60th book he has written on his own or with associates. He holds degrees from Stanford University and the University of California, Los Angeles.

OTHER BOOKS BY GODFREY HARRIS

Courting Failure (with Kennith Harris)
Timeline—The Principal Events During the Lifetime of Leonardo da Vinci
Leonardo Quotebook
The Life and Contributions of Leonardo da Vinci (with Thomas Mankowski)
Coloring with Leonardo (with Daniel Mankowski)
Catalog of the Da Vinci Experience Exhibit — 1st, 2nd, 3rd, 4th Editions
What a Great Idea!
The Definitive Southern California Diet (with Jeffrey I. Barke, M.D.)
The Hottest Ideas in Word of Mouth Advertising
The Complete Business Guide
The Essential Gift Planning Kit
The Essential Diet Planning Kit (with Jeffrey I. Barke, M.D.)
The Essential Wedding Planning Kit
The Essential Cooking Planning Kit
The Essential Project Planning Kit
Civility
Corruption
The Essential Moving Planning Kit—1st, 2nd, & 3rd Editions (with Mike H. Sarbakhsh)
The Essential Travel Planning Kit—1st & 2nd Editions
Grandparenting
The Essential Event Planning Kit—1st, 2nd, 3rd, 4th, 5th, 6th, 7th, 8th, & 9th *Editions*
Watch It!
Concentration—1st & 2nd Editions (with Kennith L Harris)
Let Your Fingers Do the Talking
Talk Is Easy
The Ultimate Black Book—3rd Edition (with Kennith L Harris and Mark B Harris)
Don't Take Our Word for It!
How to Generate Word of Mouth Advertising (with Gregrey J Harris)
Promoting International Tourism—1st & 2nd Editions (with Kenneth M. Katz)
European Union Almanac—1st & 2nd Editions (with Hans J. Groll and Adelheid Hasenknopf)
The Panamanian Problem (with Guillermo de St. Malo A.)
Mapping Russia and Its Neighbors (with Sergei A. Diakonov)
Power Buying (with Gregrey J Harris)
Talk Is Cheap (with Gregrey J Harris)
The Fascination of Ivory
Invasion (with David S. Behar)
The Ultimate Black Book—2nd Edition (with Kennith L Harris)
The Ultimate Black Book
The Panamanian Perspective
Commercial Translations (with Charles Sonabend)
From Trash to Treasure (with Barbara DeKovner-Mayer)
Panama's Position
The Quest for Foreign Affairs Officers (with Francis Fielder)
The History of Sandy Hook, New Jersey
Outline of Social Sciences
Outline of Western Civilization